SLUMS AND SLUM CLEARANCE IN VICTORIAN LONDON

HISTORY OF THE CITY

SLUMS AND SLUM CLEARANCE IN VICTORIAN LONDON

J.A. YELLING

Routledge
Taylor & Francis Group

LONDON AND NEW YORK

First published in 1986

This edition published in 2007
Routledge
2 Park Square, Milton Park, Abingdon, Oxfordshire OX14 4RN
711 Third Avenue, New York, NY 10017

First issued in paperback 2014

Routledge is an imprint of Taylor & Francis Group, an informa business

British Library Cataloguing in Publication Data
A CIP catalogue record for this book
is available from the British Library

Slums and Slum Clearance in Victorian London
ISBN10: 0-415-41816-X (volume)
ISBN10: 0-415-41933-6 (subset)
ISBN10: 0-415-41318-4 (set)

ISBN13: 978-0-415-41816-4 (volume)
ISBN13: 978-0-415-41933-8 (subset)
ISBN13: 978-0-415-41318-3 (set)

ISBN 13: 978-1-138-87402-2 (pbk)

Routledge Library Editions: The City

SLUMS AND SLUM CLEARANCE IN VICTORIAN LONDON

J. A. Yelling

Birkbeck College, University of London

London
ALLEN & UNWIN
Boston   Sydney

Allen & Unwin (Publishers) Ltd,
40 Museum Street, London WC1A 1LU, UK

Allen & Unwin (Publishers) Ltd,
Park Lane, Hemel Hempstead, Herts HP2 4TE, UK

Allen & Unwin, Inc.,
8 Winchester Place, Winchester, Mass. 01890, USA

Allen & Unwin (Australia) Ltd,
8 Napier Street, North Sydney, NSW 2060, Australia

First published in 1986

British Library Cataloguing in Publication Data

Yelling, J. A.
    Slums and slum clearance in Victorian London.——(London research
series in geography. ISSN 0261–0485; no. 10)
1. Slums——England——London——History——19th century
I. Title   II. Series
307'.3364'09421   HV4086.L6
ISBN 0–04–942192–1

Library of Congress Cataloging in Publication Data

Yelling, J. A. (James Alfred), 1940–
    Slums and slum clearance in Victorian London
(The London research series in geography; 10)
Bibliography: p.
Includes index
1. Slums——England——London——History——19th century
2. Poor——England——London——History——19th century
3. Urban renewal——England——London——History——19th century
I. Title.   II. Series
HV4085.L6Y45   1986   307.3'364'09421   86–3498
ISBN 0–04–942192–1 (alk. paper)

Set in 10 on 11 point Bembo by Bedford Typesetters Ltd, Bedford,
and printed in Great Britain by Anchor Brendon Ltd.,
Tiptree, Essex

Preface

This book is the product of a switch in the direction of my research. Previously, I had worked for many years on agricultural topics, this culminating in a book entitled *Common field and enclosure in England 1450–1850*, published in 1977 by Macmillan. The parallels between common fields and enclosures and slums and slum clearance should not be pushed very far; nonetheless, there are a few links between my two studies in both content and method. Enclosures and slum clearance both involve the transformation of a local geography, but their significance reaches out to embrace such very large and related themes as the nature of land and property ownership, the economic workings of the market, and the nature of society. The advantages of studying slums and slum clearances together are somewhat similar to those of studying common fields together with enclosures. Above all, I have been concerned to show that the geography of such operations as enclosure and slum clearance is not to be set aside as a matter of establishing local detail after the economic, political, and social principles of such developments have been determined. On the contrary, the study of these geographies needs to be an integral part of the effort to establish such principles on firmer and more precise foundations.

In thanking those who helped in the preparation of this book, my thoughts go first to my wife Geneviève and my daughter Emily, who supported me through all the difficult periods. Thanks are also due to Loraine Rutt, who drew the figures under the supervision of George Reeve. I am also grateful to Gwyn Meirion-Jones and Philip Stott for their assistance in the final stages. The Greater London Record Office kindly gave permission to use its plans in the compilation of Figures 5.1 and 7.1. I should like to thank the staff of the Record Office for their consistently courteous attention over many years.

J. A. YELLING

Contents

List of figures

List of tables

1 Introduction

Victorian London is a classic site of the slum. Its notorious presence in the capital of the world's most economically advanced country stimulated the first discussions of that since recurrent problem of 'poverty in the midst of plenty'. Projects formed in response to that presence, ranging from redevelopment to suburban municipal estates and Ebenezer Howard's new towns, were themselves to become internationally important landmarks in the development of modern urbanism.

Not surprisingly, these events and conditions have already been the subject of notable modern works by Dyos, Stedman-Jones, Wohl, and others.[1] My own point of entry into this much discussed world is made through a closer study of the process of slum clearance. It covers the development of policies and programmes from their initiation through Cross's Act (1875) to the abandonment of clearance by the London County Council at the end of the Victorian period in favour of a suburban solution. It is concerned with the manner in which such policies related to the nature of the slum and its place in the urban structure. The discussion ranges from contemporary understanding of such matters to the detailed content and repercussions of policies, which required the designation of unfit houses, the compensation of property owners, the displacement of tenants, and the rebuilding of sites.

London forms a favourable context for such a study in that it allows the focus of investigation to be shifted easily from national policies through to the operations of the executive authorities at the whole London level, and to the detailed analysis of particular areas and clearance operations. In the Victorian period, government policy regarding the slum was shaped mainly in relation to the capital, and the two principal pieces of housing legislation, the Acts of 1875 and 1890, notably reflect this. Social writers of the period also gave principal attention to London, and the work of Charles Booth in *Life and Labour of the People in London* is especially significant for this book.[2] Victorian London was a city close enough at hand for reference to contemporary conditions to be made in some detail, yet it possessed extensive scale and varied urban structure. On the one hand, notorious slums reflected the intense particularity of its various districts; on the other hand, slums and slum clearance were also related to a larger geographical organization reflected in land values, workplaces, and transport.

'Slum' is here treated as a term in the discourse of politics rather than science. It carries a condemnation of existing conditions and, implicitly at least, a call for action. Such action need not, of course, involve slum clearance, and even if it does that clearance may vary enormously in form and meaning. The reason for this variety lies in the range of conditions to which the term 'slum' may be applied – both in nature and degree – and also in the range of remedies that can be offered. The possible combinations of problem and remedy soon build, so that even at a given time and place there is no uniform view. However, for given places there exists in most periods a paradigm strategy which incorporates dominant views of the nature of the slum, and of the remedies which should be applied. Cross's Act embodied such a strategy for the late Victorian Age.

It should be re-emphasized that the term 'paradigm' is here applied to a political not a scientific activity. It is useful in underlining the fact that a strategy involves a way of seeing the slum, of selecting and emphasizing some aspects and not others. It is a problem-selecting as well as a problem-solving device. Moreover, as a relatively concrete entity it is necessarily built up from a combination of ideological and empirical referents. These embrace not only the nature of the slum but also the nature of the proposed remedies. For if the view of the slum shapes the remedies, it is also true that the nature of the remedies shapes the view of the slum. Whether in the general approach to the slum or in the detailed complexity of, say, the compensation question, this relationship is central to policy outcomes.

Such a strategy is thus a complex; it cannot be explained by tracing any single line of derivation either from ideology or from urban structure. Its plausibility depends on success in drawing certain key strands of argument together, so that they may be seen to fit into a coherent framework. A concentration on this 'fit', and on unravelling the strands thus woven together in the rationale of the strategy, should enable a deeper penetration into the thought and action concerning Victorian slums and slum clearance. For from the central nub of strategy it is possible to connect on the one hand with more abstract frameworks of political and economic philosophy and on the other with the structure and workings of the urban system.

The term 'slum clearance' is used here in its official sense, and therefore signifies government intervention. As such, an ideology which stuck rigidly to non-intervention could not incorporate it, and this was, of course, especially significant in the Victorian period. At a very general level, this period was characterized by attempts to deal with social problems thrown up by the distributional workings of a relatively pure market economy, coupled with an equal stress on the avoidance of any intervention which involved transfer of resources from one social group to another. Such an equation made any straightforward progress unlikely, and although policies might be justified on pragmatic grounds, they were shaped by ideological considerations which drew greater attention to some matters than others, and made it easier to part in certain directions than others. Slum clearance was one of these favoured directions. The Victorian period was, indeed, marked by its attention to what Booth called 'the quagmire underlying the social structure'.[3] Social and environmental deterioration might be arrested and reversed by taking action at the base, where conditions were most easily presented as special cases requiring special treatment. Hence, the slum might be regarded as a moral and sanitary problem, and those within its borders, both tenants and landlords, depicted as distinct from their counterparts in the rest of society.

Political strategies such as that embodied in Cross's Act are not, however, the result of any profound reworking of empirical referents according to clear-cut ideological criteria. Instead, they are eclectic, covering the variety of points to which any concrete strategy must necessarily refer by drawing on existing strands of thought and action, each with its own history. The first function of a strategy is persuasion, without which there would be no adoption, and for this a certain coarseness is an advantage, reconciling different interests, and enabling difficulties to be glossed over. Particular attention is concentrated on the defects of existing arrangements. Thus, although strategies were built up over a period of time, and were not simply opportunistic, the presence of logical weaknesses or inadequacy of empirical foundations does not prevent their advocacy or acceptance. Such weaknesses and inadequacies are inevitably incorporated into any strategy, and in the case of Cross's Act were of a serious nature, and immediately revealed on implementation.

The experience of implementation thus provides points around which counter-strategies may be formed. Contradictions are revealed, whereas success itself may diminish the apparent need to continue in the same direction. The notion of paradigm is again useful here in that it underlines the point that damaging experiences do not necessarily give rise to any immediate abandonment of an existing strategy. It may simply be modified, or changed in its scale of application as expectations are lowered. Scope exists for such modification because of the relatively loose binding of the parts, and this is an important aspect of strategies. Thus, although initial enthusiasm will have dulled as difficulties are encountered, the option may well be to carry on within the existing strategy unless an attractive alternative is available. I shall argue that this is what happened with Cross's Act. For although rooted in the experience of the previous strategy, an alternative does not gain acceptance simply by direct assault on propositions previously accepted. It also relies very much on a switch of attention, on providing other interests and objectives. If policy makers at the turn of the century had been confined to the same view of 'slum' as their mid-Victorian predecessors, they would have prescribed much the same kind of remedy.

A suburban solution thus came to prominence not just because of the failure of previous remedies but because of an emphasis on new aspects of 'slum' for which different remedies seemed to be appropriate. Indeed, I shall argue that it involved a distinctly different conception of the historico-geographical development of the slum. At the same time, public intervention directed towards urban expansion implied less concern with discipline and control. Characteristically, new paradigms take some time to build up, but they win acceptance in a relatively short period of time as one way of seeing things becomes replaced by another. The acceptance of a suburban solution at the end of the Victorian period fell very much into this pattern. Although events at that time may be important in aiding the final breakthrough, the replacement of one strategy by another obviously requires a wider form of explanation, in both political and technical directions.

Because of the manner in which policies developed, the new strategy cannot be regarded as a straightforward advance onto new ground. For one thing, it was not completely new, but involved a reworking of old and new elements into a distinctive and original combination. For another, the factors which the old strategy had attempted to grapple with had been bypassed rather than overcome. Much the same applies in detail to the developments and shifts of emphasis which occurred within strategies. Any advance in one direction frequently threw up difficulties in another. At best, forward movement could be achieved only in a crablike manner, and relatively slowly.

In the first part of the book the emphasis is on policies and programmes. In Chapter 2 the concern is with the rationale of Cross's Act and the manner in which it was established as a paradigm strategy. That chapter finishes with a section concerning the initial difficulties of implementation. In the third chapter it is shown how the strategy was then recast in a less adventurous form, and a narrative of events is given, culminating in the eventual abandonment of clearance by the London Progressives in favour of a suburban solution. Then, in the final chapter in Part I, I review these developments and examine how the alternative rationale of a suburban solution was established. Special attention is given to the writings of Charles Booth and William Thompson.

The concern in Parts II and III is with two key areas in the application of slum clearance programmes. Slum property needs to be identified and separated from other types. Rules have to be drawn up for compensation which must effect the cost of schemes, the designation of property, the level of payment to different types of proprietor, and thus ultimately the politics of clearance itself. Similarly, the characteristics of tenants affected by clearance schemes need to be considered in relation to the objectives of schemes, rehousing operations need to be assessed in relation to costs, rents, and building standards, and the fate of the tenants and the effect of schemes on subsequent slum formation needs to be examined.

Although the focus of interest in Part I is on policies and programmes, in the second and third Parts I strike more deeply into the contemporary structure and workings of urban economy and society. Certainly, most detailed argument involving manipulation of data is to be found in these latter parts. However, the concern is not solely with the empirical side of the implementation of policies. Some aspects of policy which require lengthier treatment are also dealt with there, and above all the aim is to examine how policy and empirical referents fitted together. Thus, although the predominant movement of argument is from Part I to Parts II and III, the latter sections are also intended to add to the discussion of the rationale and development of strategies with which the book begins.

Archival sources

Only a few Home Office papers survive at the Public Record Office (PRO), so that the following account depends particularly on the archives of the Metropolitan Board of Works (MBW) and the London County Council (LCC), both housed at the Greater London Record Office. In addition to the printed minutes of the Board and the Council, minutes and presented papers are available for the committees which handled housing operations. For the Board this was the Works and General Purposes Committee (WGP), and for the Council the Housing of the Working Classes Committee (HC), although this name undergoes some variation. In addition, the Board's records include reports of Home Office local inquiries into schemes, and the awards of arbitrators, as well as a series of printed reports bearing on general policy. The Council's dealings with compensation are mainly documented in the HC Presented Papers, and it also made a collection of more general housing papers (HSG). Outside of these official archives, one source that should be mentioned is the Booth Collection at the British Library of Political and Economic Science, which includes the unpublished notebooks of the Booth survey.

Notes

1 Dyos, H. J. 1982. The slums of Victorian London. In *Exploring the urban past*, D. Cannadine and D. Reeder (eds), pp. 129–54. Cambridge: Cambridge University Press; Dyos, H. J. and Reeder, D. 1973. Slums and suburbs. In *The Victorian city*, H. J. Dyos and M. Wolff (eds), vol. 1, pp. 359–86. London: Routledge & Kegan Paul; Jones, G. S. 1971. *Outcast London: a study of relationships between classes in Victorian society*. London: Macmillan; Wohl, A. S. 1977. *The eternal slum: housing and social policy in Victorian London*. London: Edward Arnold. For other works relating particularly to rehousing operations see Chapter 8.
2 Booth, C. 1889–1903. *The life and labour of the people in London*. 17 vols. London: Macmillan.
3 ibid., 1902–3 edn, vol. 1, p. 176.

Part I

POLICIES AND PROGRAMMES

2   A new policy against the slum

The Artisans' and Labourers' Dwellings Improvement Act introduced by Cross in 1875 aimed to remove slums and put new working-class dwellings in their place. The midwife for this operation was to be the local authority, which would designate the site, compensate the owners, clear the land, and sell it to private developers. One geography would be replaced by another. Needless to say this programme, simple in outline and modern in conception, bristled with difficulties of all sorts. These arose in detail in the practical application of each stage of the operation. More fundamentally, they reflected problems in aligning the desired reconstruction with prevailing economic and social structures and processes. Despite this, Cross's Act, modified in detail, was to form a persistent framework for clearance in our period, and the manner of its introduction warrants close attention.

A necessary precondition was the existence of a relatively favourable political climate for the reception of such proposals. Disraeli, whose Conservative government took office in February 1874, had seen the need to extend his party's rural base into the towns, and to widen the social range of its appeal in the light of the extension of the franchise in 1867.[1] He accepted a link between social reform and sanitary improvement and extended the latter to include housing. Sanitary policy could be presented as beneficial to all, and it was a now traditional field of state intervention to which new initiatives could be attached. Introducing his Bill, Cross, the new Home Secretary, was to stick closely to this point. 'It is not', he said, 'the duty of the Government to provide any class of citizen with any of the necessities of life... [but]... no one will doubt the propriety and right of the State to interfere in matters relating to sanitary laws.'[2]

Disraeli, however, had little else to offer, and any practical proposals had to come from elsewhere. In this case the chief source was a Dwellings Committee consisting of members of Parliament, representatives of dwellings companies, medical officers of health and others, which had met under the auspices of the Charity Organisation Society (COS) and reported in November 1873. They reached a substantial measure of agreement on a programme needed to deal with slum conditions in London, which one of their members, Dr Liddle, summarized as follows:

> The best plan for improving these densely crowded localities... is for the Metropolitan Board of Works to obtain powers for the compulsory purchase of lands and houses which are unfit for habitation, and sell the ground... for the purpose of erecting suitable houses for... the working classes.[3]

These proposals, buttressed by a memorial from the Royal College of Physicians of London, were presented to Parliament by two Liberal members of the Committee, Kay-Shuttleworth and Waterlow, in May 1874. In the debate Cross promised legislation in line with the spirit of the representations.

The logic of Cross's Act: general points

Housing had entered the sphere of sanitary policy more forcefully in the 1860s. Official slum clearance then began through local Improvement Acts obtained in Liverpool (1864), Glasgow (1866), and Edinburgh (1867). A defining characteristic of such clearance is that it stems from the representation of a medical officer of health. It is this which distinguishes it, in the technical sense, from other actions, both public and private, which involve the demolition of slum housing. Torrens's Act (1868) and Cross's Act itself were designed to extend this new kind of clearance to other cities, and above all to the metropolis.

In defending his measure, Cross was at pains to emphasize its scientific basis: 'The plague spots where all these evils flourish and whence they spread . . . are mapped out . . . as clearly as the mountains of the moon are, by the aid of scientific discovery.'[4] Such insistence, however, reflected a concern with the political legitimization of clearance as much as a confidence in the ability of medical science to provide an unchallengeable technical means of recognizing the slum. For already the actions begun in provincial cities took very different forms. In Liverpool clearance was scattered and very selective, whereas in Glasgow the Improvement Act took in 88 acres, 'including practically the whole of what was Glasgow in Adam Smith's time'.[5] And the 'plague spots' which were to be cleared in London under Cross's Act differed in scale and form from both of these examples.

Such distinctions may in part reflect objective differences in slum conditions from one city to another, or differing perceptions of medical officers regarding the nature and intensity of the slum problem. But they also stemmed from differences in the nature of the remedies proposed. Clearance programmes might thus be confined to demolition, or include rehousing, and they might range from minor alterations to projects which reorientated the whole geography of a city. These matters reflected a political choice well beyond the competence of medical officers to determine, and they imposed their own constraints on the designation of clearance areas. Clearance policies and programmes have, therefore, to be treated as a package, for if the perceived nature of the slum is the point from which all action stems, it is nonetheless true that the proposed future of the cleared land and of the occupants of the demolished buildings in turn shapes the definition of 'slum'.

It is this necessarily political nature of slum clearance which requires that it be considered not just as a technical operation but in relation to the social and political thought and action specific to particular places and periods. Concerning Glasgow, for example, Tarn rightly emphasizes that the 1866 Act 'made its redevelopment possible on a scale only comparable with a town planning scheme today'.[6] However, in the letter which Gairdner, MOH for Glasgow, wrote to the COS Dwellings Committee to emphasize the importance of 'the destructive part of the duty of the authorities', the operation of clearance loses its air of modernity, and is plunged back firmly into the world of mid-Victorian Glasgow:

The great demolitions by railway companies often work in the same direction in the end, but not being guided by the clear principle of demolishing gradually and beginning with the worst, these destructions have been attended with great and unmerited suffering to the respectable poor, and have given a bad name to what is nevertheless the only true process of reformation for our towns, viz. to make them in time literally uninhabitable for those dangerous classes who at present live in them by sheer force of chronic neglect, and with a sort of vested interest in every kind of sanitary, moral and religious degradation.[7]

In describing the motives which underlay the Glasgow improvement, Allan gives some part to philanthropy, more to 'fear of the reservoirs of disease, crime and vice', and most to 'the feeling that slum areas were an affront to civic dignity'.[8] Smith, in his account of the Edinburgh scheme, concurs.[9] Such motives, in varying proportions, recur in the pressure by London groups to obtain Cross's Act, and in the subsequent history of its application. At the outset, however, attention may be given to other considerations which cut across this kind of attribution and relate more directly to the moves which could be made on the political chessboard. Their effect was to demarcate the slum as a special problem requiring special treatment.

There was a distinct tendency for Victorian political discussions to weave around the complementary notions of 'laxity' and 'responsibility'. This derived in part from an insistence on the validity of existing structures of political economy, so that perceived defects in social conditions were related to agents rather than structures, to administration rather than law. It was closely connected also to the high moral tone of much political debate. Although these ideas had general application, they were particularly relevant to symptoms of chronic failure, as represented in the slum. This was above all seen as a product of laxity, of unwise neglect and failure of duty. The antidote was an assumption of responsibility, and here a whole hierarchy of duties might be envisaged ranging upwards from those of the tenant to middleman, ground landlord and local and central government. All these agents were bound in a chain of responsibility, even though it was an unequal one. For Booth, 'the owner of a house, if he be given and if he accept his proper place, will be but the medium by which punishment will fall on the evil doer and order be enforced, and by the performance of this service he will acquire a new and noble title to his property'.[10]

If, however, unacceptable conditions persisted, then this was frequently seen as a failure of will, which in turn might be explained by a variety of factors ranging from weakness of resolve to the pursuit of vested interests and in which tenants, landlords, and public authorities might all be involved. The remarks of Sir John Simon are representative:

When housing infamously unfit is permitted to tender itself for hire, and when the laws which have been enacted against nuisances, and against . . . overcrowding . . . are by negligence or corruption left unenforced . . . the real compassion has not been for the earnings of the poor, but for the profits of the house jobber or landlord.[11]

This type of discourse was thus not wholly reactionary in its implications. It was easy to slip from an attack on individual slum landlords through to an attack on landlords as a class. An emphasis on inefficiency or corruption in local government could well lead on to a recognition of the need for administrative reform. At the top, the responsibility of central government for social deficiencies could not easily be strictly confined. According to T. H. Marshall, 'If capitalist industry was one of the pillars upholding the Victorian system, the other was responsible government, both central and local.'[12] The great question was whether these pillars could be held together.

In the mid-Victorian period, however, attitudes to poverty were shaped by the constant assertion that pauperism was promoted by laxity in the administration of the Poor Law. The 'Speenhamland system', operated prior to the reform of 1834, was the special symbol of dependence fostered by public relief, but subsequently the principles of 1834 had not always been rigorously enforced. The Charity Organisation Society, founded in London in 1869, was particularly concerned to promote a return to those principles, and to stress that dependence on relief, whether public or private, must inevitably weaken self-reliance and individual effort. This was all the more important in that the working class, through their votes, might soon be in control of the public purse. The formation of the COS was followed by a strengthening of resolve to press higher standards on the poor through greater self-discipline. The Society supported the Poor Law inspectors in the 'crusade against outdoor relief', which began in the early 1870s and lasted into the 1890s.[13] A crusade against the slum was a natural concomitant.

The writings of John Simon illustrate the manner in which such attitudes could be linked to sanitary reform. He accepted the doctrines of the COS in relation to relief, believing that it was 'among the first conditions of good government that the community shall sharply distinguish between those of its body who are self-supporting . . . and those who more or less depend on support from public alms'. But alongside, or perhaps underlying, the laws of political economy, there were sanitary laws which needed equal enforcement. Indeed, just as insistence on the laws of political economy was necessary for the true advancement of the respectable poor, 'efficient administration of the sanitary laws is among the best helps that can be given to the poorer classes'. Sanitary laws would protect the poor against the temptation of living cheaply in unfit housing, just as they were already protected against diseased or rotten food. However,

> in the supply of dwellings as in the supply of food, for the self supporting labourer, when once the conditions of qualitative fitness are duly secured by law and administration, facts of quantitative insufficiency are for ordinary commercial enterprise to meet.[14]

Even in the mid-Victorian period, this distinction between qualitative and quantitative was difficult to maintain, and the 1875 Act itself involved a major degree of compromise, despite the language with which Cross introduced the measure. He therefore made full use of another important way in which the slum might be set apart. This came from an insistence that the slum was the product of the past. So Cross could say, 'When these slums, the accumulation of

centuries, have been cleared away, unless there be great neglect in the administration of the ordinary law, and new slums be allowed in consequence to grow
up, this Act will have done its work.[15] This implied much more than an assertion
that a finite stock of housing no longer suitable for present purposes was being
removed. It implied also that the slum was the product of a past society. There
was no necessary reason why slums should exist in the current economic and
social system, given efficient administration. Similarly, although slum conditions might spread from the 'plague spots' in which they were anchored, the
existence of these plague spots was wholly contingent, that is they occupied no
necessary place in the modern city.

There was some analogy in Cross's legislation between slum clearance and
the operation of building bylaws which were being concurrently pressed in the
Public Health Act of 1875. Such bylaws were designed, *inter alia*, to reduce
public health problems in new developments by requiring greater provision of
space around houses for access of light and air and location of essential
conveniences. Medical officers under Cross's Act were also to designate sites
where houses were 'unfit for human habitation' because 'closeness, narrowness,
and bad arrangement' resulted in 'want of light, air, ventilation and proper
conveniences'. The Act could therefore be seen to involve the removal of a
bottom tier of housing which had not been controlled by modern sanitary
legislation.

These matters also had political importance in their bearing on the crucial
question of property compensation. For this was one of the key points at which
the perceptions of the slum came into contact with the mechanisms at work in
slum formation. One line of thought, stressing the responsibility of the individual for his property, opened up the possibility that the owners of slum houses
might be treated in the same manner as the dealers in rotten food. Such
principles were incorporated to some extent in Torrens's Act, which placed first
responsibility on the owner, the local authority acting only in default. Correspondingly, the measure was framed in relation to individual houses, and
involved the notion of rehabilitation through which the owner could demonstrate the resumption of his responsibilities. Demolition was only a last resort,
but in this case there was no compensation for the demolished building.

In Cross's Act the responsibility of the owner for the maintenance of his
property was largely set aside by emphasizing those aspects of the slum over
which he currently had no control: the 'closeness, narrowness, and bad arrangement' of the buildings. The Act began from the position that as slums were 'the
property of several owners, it is not in the power of any one owner to make such
alterations as are necessary for the public health'. In addition, there was the
argument, used for example by Octavia Hill, that since the state had failed to lay
down sanitary regulations to prevent the disorderly growth of slums, these
represented the 'dulled conscience' of the 'community', and one could not 'fairly
throw the cost on the individual owner'.[16] Compensation should thus be based
on market values.

It was at this point that local authority intervention was required. For
although this was justified by the need for compulsory purchase of property, it
was also clear that generous compensation meant greater expense. The rôle of
the local authority would therefore also be to absorb the costs of removing the

'mistakes of the past'. Once this had been accomplished, however, development was to be carried out by private bodies in order to ensure its strict accordance with commercial principles. For, as the Dwellings Committee put it, 'philanthropic agency in building dwellings for the poor means the supply of one of the chief necessities of life . . . below its market value. Were such a principle to be . . . sanctioned, not only would the profits of commercial investment be impaired, but the principle of self dependence would be attacked.'[17]

The duty of the local authority under the Act was thus to prepare the site, to impose conditions on 'the elevation, size, and design of the houses and the extent of the accommodation', and to make 'due provision for the maintenance of proper sanitary arrangements'. It was required that each scheme should normally provide 'for the accommodation of at least as many persons of the working class as may be displaced . . . in suitable dwellings . . . within the limits of the same area or the vicinity thereof'.

This last point was to be a key provision of Cross's Act, for the complementarity between its destructive and reconstructive parts was a feature which specially distinguished it from other types of clearance legislation. Cross's Act has to be understood as a measure designed to provide housing sites as well as to demolish slums, and it was around the accordance of these two objectives that the logical structure of the Act was built. This accordance imposed a certain constraint on the designation of the slum, for it required that slums be identified that were suitable in shape and size for the work of rehousing agencies. Without such rehousing provisions it was possible to engage in more sporadic destruction, as in Liverpoool, or a very large scale attack on the existing fabric of the city, as in Edinburgh or Glasgow. The tight relationship between destruction and reconstruction in Cross's Act also required the existence of specific agencies that might be entrusted with rebuilding, and it presupposed that the necessity for such rehousing had already been clearly established. It was here that the metropolitan origins of the Act showed most clearly.

Cross's Act in its London context

The Scottish Improvement schemes of the 1860s stemmed, Smith says, from earlier improvements in which slum clearance formed part of a variety of objectives dominated by street formation. It was a logical step to reverse the relationship between street improvement and slum clearance and 'to design improvement schemes which had slum clearance as their first objective'.[18] In early Victorian London, too, Dyos has shown that the objects of street improvement 'were seldom single, for street improvement during these years provided almost the only effective way of rectifying on a grand scale some of the worst features of urban growth'.[19] However, it soon emerged that attempts to disperse slums in this way were not wholly successful. The New Oxford Street improvement showed this most clearly. The Select Committee of 1838, which approved the scheme, believed that 'the most important improvements' would be 'amendment of health and morals' by 'the removal of congregations of vice and misery and the introduction of a better police'.[20] A decade later, however, overcrowding in St Giles could be ascribed to 'what are falsely called the improvements which have recently taken place... in the formation of New Oxford Street ... The expelled inhabitants are forced to invade the yet remaining hovels suited to their means'.[21]

Demolitions for street improvement were soon accompanied by larger-scale clearances for railway operations. Even more than the new streets, the introduction of railways seemed paradoxical in its effects. The great technical achievement which was supposed to stimulate the suburbs and liberate the industrial city of its congestion had the immediate effect of increasing overcrowding in the areas surrounding the demolitions. Reformers such as Shaftesbury attempted to slow down these developments and to make them more subject to parliamentary control. The controversy was particularly sharp in the period 1859–62 when intense railway activity caused the displacement of nearly 37000 people.[22] *The Times* was not slow to point out the irony of the situation: 'That Lord Derby should call upon the legislative to interfere to protect these fever preserves and these crime coverts is indeed strange. Stranger still... to hear Lord Shaftesbury bemoaning the destruction of that dreadful rookery by Field Lane.'[23]

The Times believed that it was both convenient and necessary that the habitations of the labourer should become more dispersed. Most reformers, however, drew the opposite conclusion. It was not economically possible for the labourer to live in the suburbs; he had to be housed in the centre adjacent to his means of livelihood. Already in 1853 Shaftesbury had unsuccessfully proposed that a rehousing obligation be attached to parliamentary bills promoting railway schemes which required demolitions. By the early 1870s opinion had become more favourable. In 1872 the Metropolitan Street Improvements Act had included a provision requiring some rehousing. In 1874 Cross himself introduced new standing orders aimed at controlling railway demolitions, and requiring the provision of alternative housing.

The presence of Lord Shaftesbury on the COS Dwellings Committee meant that these considerations would be powerfully presented. Not surprisingly, therefore, the first conclusion of the Committee was that 'the mass of the poorer classes must be provided with habitations near their work'. Suburban cottages offered many advantages and their inhabitants deserved 'every facility of locomotion which the railway companies can afford to grant'. It was useless, however, to look in that direction for any immediate solution of housing problems.[24] This view was also overwhelmingly accepted in the parliamentary debate on Cross's Act. Only a few lone voices, like Lord Montagu's, argued that local authorities should be empowered 'to take sites compulsorily outside a town', and that if this were done 'rents would be lowered, dwellings would be more healthy, towns would be less crowded, and many of the working classes would have gardens and the sight of green fields'.[25]

Alongside these developments, and stimulated by them, the growth of model dwellings companies was another major strand in the chain of circumstances leading to Cross's Act. These companies had been specifically brought into being to supply new working-class housing in the central districts of London through economical tenement block construction. The earliest, such as the Metropolitan Association for Improving the Dwellings of the Industrious Classes (MADIC), dated back to the early 1840s; but the early 1860s had seen a major stimulus with the foundation of the Peabody Trust in 1862 and the Improved Industrial Dwellings Company (IIDC) in 1863. Despite this, the Dwellings Committee found that the efforts of the companies had failed to deal with the slum problem. On the contrary there was 'convincing evidence that in some parts the evil is positively progressive, while in most the antiquated character... and the decayed condition of the habitations abandoned to the poor, threaten to perpetuate the depravity, degeneracy, and pauperism which at present weigh so heavily upon the community'.[26]

The response of the companies to these criticisms was to be most important in relation to Cross's Act. Among their representatives on the Committee Sir Sidney Waterlow, chairman of the IIDC, was especially vigorous in maintaining that the companies were willing and able to supply new working-class accommodation. He claimed: 'The chief, indeed I may say the only, difficulty in the way of rapid progress of this work of private individuals is the impossibility of obtaining sites in densely populated districts.'[27] Equally important was the rationale provided for this 'inability to purchase'. It was generally acknowledged that the difficulty was partly an economic one: that the cost of acquiring such sites was a prohibitive factor. The mission of the companies, however, was to show that the housing problem could be overcome without departing from the rules of a market economy. They therefore pointed away from this issue towards the problem of site assembly. Relatively large sites were required to make housing operations efficient, and dealing with many proprietors, especially where leasehold prevailed, imposed major costs and delays. An example of such difficulties lay in the frustrated efforts of the Peabody Trust to assemble a site around land acquired from the Foundling Hospital in Coram Street, Bloomsbury. These claims of the companies were enthusiastically incorporated in the Dwellings Committee's report:

> [It was] certain that by systematic distribution, by economy of space, and greater elevation of structures, one half more people might be lodged in a comfortable and wholesome manner where the present occupants are huddled together in dirt, discomfort and disease. That the undertaking can be rendered fairly remunerative has been proved by the profits of the existing companies. The principal, almost the only obstacle to the prosecution of this great reform lies in the inability to purchase.[28]

Cross was therefore presented with a very tempting prospect. On the one hand respectable agencies seemed to be actually demanding sites on which to build working-class housing; on the other hand, there were the 'plague spots' for which there no longer seemed any appropriate means of demolition. All that was needed to bring these two together was for the state to unlock the obstacles to enterprise.

With its twin objectives of removing slums and providing housing sites for the dwellings companies, Cross's Act necessarily departed from the approach taken in Torrens's Act and in the Liverpool clearances. This involved selectivity, with close attention to the responsibility of individual proprietors and to the particularities of physical structures. Only the worst would be demolished, and rehabilitation would be stimulated by the threat of clearance. Such an approach could not supply the kind of housing site which the dwellings companies were seeking. Under Cross's legislation attention was thus switched to areas and to features of the plan and arrangement over which the individual proprietor now had little control. There could be no account taken of rehabilitation, and slum property was seen to lie in blocks rather than in any complicated scatter which reflected individual attention.

It was notorious that Torrens's legislation had not been implemented in London. It provided a stark choice between enforcement on the one hand and confiscation and dehousing on the other. Much contemporary opinion, however, saw behind this inaction the familiar forces of corruption and vested interest. Under Cross the position of the medical officers was enhanced by removing the local surveyor and magistrate from the administrative process. Administrative arrangements were further strengthened in London by making the Metropolitan Board of Works the executive authority under the Act. This meant that, although clearance proposals originated locally, they could be made with the knowledge that any deficit would be met by the ratepayers of London as a whole. The Home Secretary retained important powers as confirming authority, exercising these on the report of his local inspector.

Although designed to speed clearance, these administrative arrangements also ensured that it would not form part of any more general plan. The designation of sites was firmly placed in the hands of local officials, not any central agency. Attention was focused on discrete sites, considered separately. The reason for this was that the form of reconstruction was already established. It was to consist of equivalent working-class housing. There was no need for such sites to be related one to another, nor for them to involve any general reshaping of the city. By contrast, the Glasgow and Edinburgh schemes proceeded from a general inventory of slum sites covering extensive areas. This larger spatial perspective allowed attention to be given to certain wider objectives, as in street formation. Lacking specific rehousing provisions, these schemes were more open-ended in terms of the kind of land use which would replace the old; and they involved a much longer time span of development. By large-scale displacement from the centre, the changes they brought about forced suburban growth and altered the whole form of the city.

The distinctiveness of the Cross solution thus reflected the force of the London campaign against transport demolitions and the presence in the capital of large model dwellings companies which did not exist in provincial cities. Beyond that, differences in urban structure played a part: the much greater scale of London, its larger and more complex centre, the greater distance of suburban land from the projected clearances. Wider political factors were, however, also important, for Cross certainly bore in mind the need to avoid any close comparison with the reconstruction carried out by Haussmann in Paris between 1853 and 1870. The Haussmann style represented an unacceptable concentration of arbitrary power, colossal public expenditure, far too large an assault on property, and a dangerous disturbance of the working classes. For all these reasons Cross was at pains to emphasize that his Act was not a 'general town improvement measure'. And although the example of Birmingham was to show that in provincial cities he was prepared to allow it to be used in a manner more comparable with its Scottish forerunners, this was not to be the case in London.

Although the aim of providing housing sites forced a certain perception of the slum, Cross's Act drew its principal strength, as I have argued, from the apparently corresponding requirements of its constructive and destructive parts. This correspondence was certainly based on London evidence, and on the manner in which medical and social reformers in the capital had long pointed to 'plague spots' in which moral and sanitary evils coincided and which, at least in general terms, appeared to offer the type of rehousing site the dwellings companies required. John Simon in his first report as MOH to the City in 1849 provided the initial impetus which linked medical and social concerns to closely packed housing 'absolutely unfit for human habitation', and recommended 'destruction of some considerable proportion of the court property'.[29] Two medical officers, Ross of St Giles and Liddle of Whitechapel, sat on the COS Dwellings Committee. The records of St Giles provide a particularly good example of the way in which 'notorious' slums might come to be recognized in the manner which Cross expected.[30]

Ross's predecessor, Buchanan, wrote in his report of 1858 that he had become 'more and more impressed with the importance of accurately determining the parts of our district which constitute most deaths to the register'. He remarked on the enormous differences thus revealed, distinguishing the Church Street rookery, as well as the Coram Street, Bloomsbury, and Wild Street, Drury Lane, districts which later figured prominently among the early representations to the Board of Works. At Lincoln Court, Wild Street, he reported:

The leaseholder was repeatedly summoned for his premises being over-crowded and otherwise very unhealthy, but no effect was produced. At last the leaseholder was ejected and the court was then shut up and repaired throughout at the expense of the freeholder. Some six or seven hundred people were turned out of the twenty two houses in this court in January 1859.

In 1862–3 he reported:

The angle of Bloomsbury around Coram Street has habitually a much higher mortality than the rest of the parish . . . the original arrangement of courts, in hollows below the street level . . . contribute an unhealthy condition with which the inspector has no means to deal . . . representations to this effect have been made in 1863 to the committee of the Foundling Hospital . . . This committee states that they have no power to undertake a thorough reconstruction of these courts until the expiration of the existing leases.

His successor, Ross, took up the theme in 1870: 'It is in vain to expect that our worst districts (the neighbourhood of Wild Street, Shorts Garden, Little Coram Street for example) can ever be improved . . . by dealing with separate houses.

Whole blocks must be demolished, large spaces cleared, the old soil removed and new dwellings adapted for the habitation of numerous families built upon the sites.' In 1874 he refers to the outbreak of typhoid fever in the Great Wild Street that originated in Lincoln Court: 'Lincoln Court has long been notorious as the nestling place of fever, and your Board has on numerous occasions insisted on large improvements being carried out there, but it has long been obvious that no partial measures can have any avail . . . I am glad to say that Little Coram St. and its surroundings, which have been equally conspicuous . . . are likely soon to undergo great change . . . (portions) . . . having been recently purchased by the Peabody Trust.'

When we know that the Trust itself was unable to overcome the problem of leases and acquire complete control over the area, an immediate connection is made with the chain of events leading to Cross's Act. Here are the familiar grounds on which the Act was based – high local concentrations of disease, a history of unsuccessful remedial measures, physical circumstances which made piecemeal action unsatisfactory, difficulties of achieving the necessary unity of control except through compulsory purchase. The importance of medical opinion in the thinking leading to the Act is also underlined.

To an extent, therefore, Cross was justified in believing that his Act applied to 'houses, courts and alleys within certain well-known and well-defined areas'.[31] This did not mean that medical officers would always find it easy to demarcate 'plague spots' which fitted the requirements of the reconstructive part of the Act. These difficulties will be examined later. There was, however, certainly sufficient correspondence to carry political plausibility, and the same is true if the delineation of London slums by social topographers is considered. They typically described as slums blocks which might range from small knots of houses to wider areas containing several thousand inhabitants. According to Dyos, 'slum' was in origin a slang word which had one meaning as 'a room in which low goings-on occurred. It was this last root from which the modern meaning of the word has developed to include whole houses and districts.' Cardinal Wiseman, in a famous passage published in 1850 which popularized the term, referred to 'labyrinths of lanes and courts, and alleys and slum, nests of ignorance, vice, depravity and crime' around Westminster Abbey. It was used alongside and eventually replaced other terms for houses or districts of low repute, notably 'rookery'.[32]

The classic example of a 'rookery' lay around Church Street, St Giles. Beames defines it as a relatively small block with definite borders: 'bounded by Bainbridge St., George St. and High St.' He emphasizes an isolation from outside influences and an inward turning to morally dubious sources. The rookery was 'like a honeycomb, perforated by a number of courts and blind alleys, cul-de-sacs without any outlet other than the entrance . . . In the centre of the hive was the famous thieves' public house called Rat's Castle.'[33] Engels (1844) provides a physical description which relates back to the concerns of medical officers: 'It is a disorderly collection of tall three or four storey houses with narrow, crooked, filthy streets . . . the houses are occupied from cellar to garrett, filthy within and without . . . but all this is nothing in comparison with the dwellings in the narrow courts and alleys between the streets entered by covered passages between the houses.' He also gives a typical account of social conditions:

Here live the poorest of the poor, the worst paid workers with thieves and the victims of prostitution indiscriminately huddled together, the majority Irish or of Irish connection, and those who have not yet sunk in the whirlpool of moral ruin which surrounds them, sinking daily deeper, losing daily more of their power to resist the demoralizing influence of want, filth and evil surroundings.[34]

In the 1850s and 1860s social topographers set out to demonstrate that the condition of this rookery was not unique, nor even the worst in London. The best of these commentaries are those of Beames, Godwin, and Hollingshead, to which may be added the more selective descriptions of Mayhew.[35] Such authors invited their readers to 'beat the bounds of metropolitan dirt and misery',[36] and to come with them into the courts and alleys which they named. In the course of the works most of the slum blocks which were later to be cleared under the Housing Acts were described. Particular attention was given to those in the West End, but also to the northern and eastern borders of the City through Saffron Hill to Whitecross Street, Boundary Street, Flower and Dean Street, and Rosemary Lane, Whitechapel. Further east or south, however, descriptions petered out, with certain exceptions, such as the Mint (Southwark) or Jacob's Island (Bermondsey).

It was generally admitted that there was 'a hideous sameness between St Giles, Saffron Hill, minories, and other haunts of the destitute',[37] but a variety of particular circumstances were held to account for these occurrences – sometimes a location near a cathedral or other opportunity for mendicacy, the proximity of docks and seamen, or employment in a gasworks. Although some slums were acknowledged to be of relatively recent origin, others had long been notorious. Beames stressed the manner in which districts originally well inhabited, such as Pye Street, Westminster, or the Drury Lane area, had over the course of centuries become infilled and subdivided as their original populations moved away. He thought it 'very interesting to trace the steps by which any particular district degenerated into a pauper colony . . . we may perhaps learn at what point to interpose, and when it becomes necessary to declare a house unfit for human habitation'.[38]

Whatever their origins, however, these 'plague spots' embedded in the physical and social fabric of London were points of corruption. Here disease, crime, vice, and pauperism were nourished, and from here they spread their contaminating influences. Beames maintained: 'Rookeries still survive by their very isolation, by their retention of past anomalies . . . meanwhile when rebellion recruits her forces she is fed by the denizens of these retreats.'[39] In 1854 Godwin wrote: 'If there were no courts and blind alleys there would be less immorality and physical suffering. The means of escaping from public view which they afford generate evil habits, and even when this is not the case, render personal efforts for improvement unlikely.'[40]

More than any other, the concept of 'alsatia' symbolized this association of corruption with isolation and linked it to a history of mistaken tolerance and unwise neglect. 'Whence came these rookeries?', asks Beames, 'Were they prison colonies, safety valves, so many Alsatias . . . tolerated as the least of two evils by the authorities?'[41] The usual example given in mid-Victorian literature was the 'Mint' in Southwark, whose history is described by Godwin. Once the site of a noble mansion, then used as a mint by Henry VII, the area later became divided amongst poor cottages. 'This district became the resort of lawless persons, the privilege of exemption from legal powers being claimed for it: it was an Alsatia, a sanctuary for evil, nor was any proper control obtained there until 1723.'[42]

Many believed that widespread slum conditions stemmed from a nearer period of aristocratic mismanagement to which their own activities, as responsible citizens of a new democracy, were a reaction. According to C. B. P. Bosanquet: 'Under Anne and the Georges... lawlessness becomes the chief characteristic of London life... Nearly all the chief haunts of degradation and vice in modern London were then in existence... The neighbourhood of Westminster Abbey and the Savoy, Drury Lane and St Giles, both sides of the Fleet Ditch, parts of Moorfields and Shoreditch, Ratcliff Highway and the Mint in Southwark... were in the possession of a class whose hand was against every man.'[43]

For such areas, saturated by moral and sanitary evil over the course of decades or centuries, only one remedy could be prescribed. Hollingshead says:

> I am no advocate for routing out the industrious poor from an over-crowded district to make room for stucco temples or ornamental squares ... but the recognized haunts of vice or crime want no ventilation, no enlargement, no tinkering philanthropy. They ought to be ploughed up by the roots.[44]

The debate on the Act

In the previous sections, I have tried to show that Cross's Act was by no means an arbitrary measure. It had strong roots in the thought of the period, and its framework was constructed from a consensus amongst the dominant voices in reform. This consensus, however, remained only a general one, organized for the promotion of what seemed the most acceptable line of advance. Pushed further, medical officers, social reformers, and dwellings company managers would have introduced many caveats, and their accounts of the physical and social nature of the slum, and of the effects of the proposed remedies, would have been more nuanced and more conflicting. The logic of the Act was sufficient to carry plausibility, but it was far from complete, and this was to be revealed particularly at two main points: property compensation and its impact on the cost of clearance, and rehousing policy and its impact on tenants and slum formation.

In accordance with the reasoning already examined, the original Bill envisaged compensation on the basis of 'fair market value', but condemned premises were not to receive the additional 10 per cent normally given for compulsory purchase. Although houses were to be condemned as 'unfit for human habitation', there was noticeably no mention of this in the compensation clause. It was argued in Parliament that if the houses were in a condition which the existing Sanitary Acts condemned, there should be 'a reduction from prima facie market value'. Steps were taken to strengthen the legislation in this respect. On the other hand Cawley insisted that compensation should not be fixed 'regardless of the value of the ground as a site'. He succeeded in adding to the clause the words 'and all circumstances affecting such value'.[45] Finally the compensation of condemned property was to be

> based upon the fair market value as estimated at the time of the valuation
> . . . due regard being had to the nature and then condition of the property
> and the possible duration of the buildings in their existing state, and to the
> state of repair thereof, and of all circumstances affecting such value,
> without any additional allowance in respect of the compulsory purchase.[46]

This left a considerable amount to the interpretation of arbitrators and juries, and contained revealing omissions, notably the absence of any mention of compensation for trading interests.

For Sir James Hogg, who spoke for the Board of Works in the Commons debate, 'the question of costs seemed the most important part of the Bill'. Several members believed that if the measure was successful it must also be expensive, and Waddy suggested something like a third loss on the amount invested.[47] Here again, Cross had received optimistic advice from the Dwellings Committee who were not 'under any apprehension that excessive charges need be incurred in connection with this work in the metropolis. The operation . . . would be gradual, and in some cases it might be rendered remunerative at once, and in most the redeemable charge on the rates would be modest.'[48] Waterlow later submitted privately to Cross a more definite figure:

> A rate of one penny in the pound will produce nearly £90,000 or sufficient
> to pay the interest on £2m. and redeem the principal in forty years. . . . £2m.
> would go a long way in paying simply the difference between the purchase
> money of the land and the sum recovered on its resale . . . I venture to say
> that £2m. . . . applied in the mode suggested would remove, to a very large
> extent, the houses in the metropolis at the present time unfit for human
> habitation.[49]

The removal of the slum problem from London at a cost of £2m was potentially an excellent political bargain, and although Cross was no doubt sceptical on this point, he was surely influenced by the optimism of the companies. In the debate on the Bill he kept noticeably to very general terms, saying, 'The ratepayers will be more than fully recouped in the long run, yet for a time at least the measure must be worked at some expense.' He pointed not only to a reduction in the death rate that would result but also to evidence from Edinburgh that clearance had resulted in a drop in serious crime and prostitution. No one disputed his claim that 'what the homes of the people are, the people themselves will be found to be'.[50] Certainly, he believed that the costs to the local authorities would be within bounds which could be justified politically by reference to such social benefits.

In conformity with the commercial principle the public authority had no control over the rents in the new buildings. There did, however, appear to be control over the size of the accommodation provided. This was an important matter. Octavia Hill, a member of the Dwellings Committee, had counselled that 'houses may be bought, pulled down and rebuilt, and the rooms in the new dwellings let at less than the rent which was paid in the original houses, and yet a return of 5 per cent net profit be made'. This meant, however, that one had to be thankful if one could 'secure for the same rent even one room in a new clean pure house. Do not insist on a supply of water on every floor.'[51] Shaftesbury, however, pointed out the dilemma: 'Are we to construct a vast number of single rooms in order to meet the needs of these poor people or are we to build houses according to our new sanitary requirements?' To construct single rooms for families would be to 'maintain a most indecent and immoral state of things', but not to do so would raise rents to impossible levels.[52]

Another argument mentioned by Kay-Shuttleworth was that 'many of the people who were paying 1s.6d. for a most miserable room were able to pay more rent'.[53] This point again referred back to the Dwellings Committee report, which claimed that the slum problem depended 'on the inability or unwillingness of a large proportion of the poorer classes to pay the rent requisite to obtain sufficient accommodation'. Referring to the evidence of Thomas Wright, 'himself a member of the working classes', it noted that 'he and others were of the opinion that a large proportion of the working classes could and ought to appropriate a larger part of their earnings to providing house accommodation for their families'. The report continued: 'It is however certain that so long as houses unfit for habitation are allowed to stand, they will be occupied on account of their cheapness, and it is therefore of great importance that such houses should be demolished as rapidly as possible.'[54]

Problems were, however, evident to those who examined the accommodation provided by the dwellings companies. To Waddy, 'anyone who had gone through the low dens described by the Home Secretary and had then visited the Peabody buildings must know that the two were inhabited by a very different class. The fact was that unless care was taken they would be sweeping away one class of people to provide residences for another.' Shaftesbury, among others, supported the idea that 'they should not confine themselves to sweeping away the dwellings complained of, but take powers to alter and improve them if possible'. Certainly, he did not hold out much hope for any orderly movement of population on clearance, and forecast an increase in local overcrowding.[55]

In the end most were prepared to accept that the slum dwellers would not be directly rehoused. As Kay-Shuttleworth put it, 'The people who went into the new buildings vacated their former dwellings, and these were inhabited by those who had lived in the demolished whynds and closes.' The replacement of the worst dwellings by new buildings would lead to each class taking a step upwards. According to many who supported this view the old population was in any case not suitable for the new dwellings. Playfair described them as 'not working men at all', but 'thieves, prostitutes and outdoor paupers'. Their dispersion was 'one of the greatest advantages of such a measure'. Waterlow admitted that dwellings companies like his own did not house the very poor, but through filtering 'the lowest and poorest classes were benefited in the only way in which with a due regard to the future they could be benefited'.[56]

Thus although various problems of rehousing policy were diagnosed with reasonable accuracy, this only resulted in contradictory solutions, none of them wholly satisfactory. These contradictions were not resolved either by the Dwellings Committee or by Parliament, and Cross himself made no attempt to do so. It left a further decisive area of ambiguity concerning the workings of the Act. Still, this was less important politically than the economics of schemes, and all kinds of problems might be surmounted if the Act were to give rise to sufficient reconstruction at reasonable public cost.

Implementation

Although the Metropolitan Board of Works was represented indirectly on the Dwellings Committee by Sir James Hogg, it had no real part in the making of Cross's Act, and it found itself charged with a programme of unknown dimensions, worried about costs, and confused over rehousing policy. There was no choice over implementation; the political imperative was obvious. This contrasted sharply with the position in major provincial cities where slum clearance, even after Cross's Act, began from initiatives by local administrations, and remained essentially under their control. It reflected the Board's weaker position as an indirectly elected body, and above all the close concern of the national government in the affairs of the capital.

The Board received its first official representation from Holborn in July 1875. It was followed immediately by one from Whitechapel and eleven others were to be received before the end of the year. The Holborn and Whitechapel proposals, however, became the first test cases, and provided the first indications of the reality of dealing with clearance sites. It was immediately apparent from the officers' reports that schemes would be expensive.[57] They made the point that the dwellings companies had previously selected such sites 'upon open spaces and adjoining public thoroughfares as they could secure on favourable terms'. The Board, by contrast, had no such choice, and had to compulsorily purchase the existing buildings and trades in areas designated on purely sanitary criteria.

Difficulties were particularly evident at Holborn where the represented area contained business premises whose acquisition 'would add at least 50 per cent to the cost of purchase'. These were therefore omitted. The Board, however, insisted on cutting out further parts, reducing the population to be displaced from 5500 to 3300. Although there were fewer difficulties at Whitechapel, both schemes were rejected by the Home Office, following local inquiries. Particular objection was made to the omission of property which lay embedded in the midst of the areas. As in addition the Home Secretary limited the new buildings to four storeys, the Holborn scheme area was judged 'too small and awkwardly shaped to rehouse the population displaced'.[58] Here the Board felt unable to meet the requirements, and in effect the project was shelved. This left the modified Whitechapel scheme as the sole survivor of these early skirmishes, and it went forward with an estimated net cost of £60 000.[59]

Despite the problems encountered so far, the Board was still in a reasonably co-operative mood when it came to decide the next schedule of clearance in the autumn of 1876. It had reached only the initial stages of property purchase at

Whitechapel, having decided to buy the whole compulsorily through the arbitrator, whose provisional award was not available until April 1877. There was still no indication that the land would prove difficult to dispose of, and at this stage the principal worry concerned the detailed configuration of sites. In October 1876 the Board considered a report on all outstanding representations. It adopted schemes in eleven localities, and placed a further eleven on a 'less urgent' list.[60]

In making up the list there seems to have been some attempt to choose schemes from all main parts of the capital. In the east, two sites at Goulston Street and Flower and Dean Street, Whitechapel, were combined to form one large scheme. A further large scheme lay on the northern borders of the City around Whitecross Street in St Lukes. Two central sites were included at Great Wild Street (St Giles) and Bedfordbury (St Martins in the Fields). South of the river three separate sites were chosen in St George the Martyr, Southwark, at Elizabeth Place, King Street, and Mint Street, later combined into one scheme. The list of selected sites was then completed by three smaller projects at Old Pye Street, Westminster, High Street, Islington, and Pear Tree Court, Clerkenwell.

In the detailed treatment of each site the Board showed its concern for financial considerations, making severe cuts at Bedfordbury and Wild Street. In the general make-up of the list, however, there is no indication that it attempted to avoid large or costly schemes. Indeed, most of the rejected proposals lay further from the centre than those selected and would have been cheaper to effect. There were varied reasons for selection, but relative judgement of sanitary conditions certainly ranked high. The Board's undertaking must therefore be seen as a fair commitment to make at this stage. The whole was costed at £412 287.

The autumn of 1876 represented the high point in the Board's dealings with the Act and with the Home Secretary, and the succeeding three years were to bring a progressive deterioration. The first blow came in April 1877 with the publication of the provisional award for Whitechapel (Rosemary Lane), and eventually the final award in December valued the property at £140 312 compared with the Board's estimate of £92 000.[61] It was clear that compensation costs for all the outstanding schemes would be much higher than envisaged.

The Board's approach now became distinctly cautious, but the momentum of clearance was to some extent maintained through 1877 and 1878 by a variety of special situations. The most important of these, disagreeable to the Board, came from the rehousing provisions attached to the Metropolitan Street Improvements Act of 1877. The bulk of the projected improvements lay in the West End in Charing Cross Road and Grays Inn Road where the Board did not wish to give up valuable street frontages to artisans' dwellings. As a result it began to look anew at those sites it was already being pressed to clear under the 1875 Act. Two of these were resurrected from the 1876 list, being within two miles of Grays Inn Road and Charing Cross Road respectively. At Essex Road, Islington, buildings of five storey height would accommodate 3590 persons, providing a surplus of 1840, and at Bowman's Buildings in St Marylebone similar dwellings could house a surplus of 570 to offset against the street displacements.[62]

In 1878 the Board agreed schemes with the Peabody Trust at Little Coram

Street, St Giles, and Great Peter Street, Westminster. These were very much within the spirit of the Act, the compulsory powers of the Board being used to acquire interests which the Trust had been unable to purchase, whereas the Peabody Trust was to pay the costs of compensation for the land which it already held as ground landlord. In addition, the Trust agreed to make surplus accommodation available for the West End street rehousing.[63] A further scheme was also adopted from the 1876 'less urgent' list at Wells Street, Poplar, which like the others was expected to provide surplus rehousing.

The next important developments concerned the disposal of the sites. The eastern section at Whitechapel was cleared and ready to be sold in July 1878, and the Board considered the conditions that should be attached to such sales. It wished to retain control over the plotting of the buildings and the division into rooms and tenements. In December 1878 the Whitechapel land was put to tender but no offers were received. Gatcliff on behalf of MADIC and Waterlow for the IIDC declined to tender, stating that the covenants were too restrictive. Peabody explained that they had not been particularly deterred by the conditions, but had 'in part a want of sufficient funds and in part a preference for other sites which the Board will have presently to offer'.[64]

The Board, in consultation with the Home Office, made certain modifications to the covenants. In February 1879 Cross wrote privately to the heads of the dwellings companies asking them to tender for the Whitechapel land.[65] The revised conditions of sale were advertised and tenders were to be opened in March. Again there were none, and it was decided to put up the land for public auction in June. In May a letter was received from the Peabody Trust offering to purchase the freehold of six sites – the eastern part of Whitechapel for £10 000 and Bedfordbury, Great Wild Street, Pear Tree Court, Whitecross Street, and Old Pye Street at a rate of 3d. a square foot on 20 years' purchase. No satisfactory offer having been made at the June auction, the Board decided to open negotiations with Peabody on the basis of 4d. a foot and 22 years' purchase. The Trust, however, declined to modify its terms and eventually its offer was accepted.[66]

This was a crucial decision, for it finally brought home to the Board the cost of the schemes which it had undertaken. One part of the equation, the compensation costs, had already proved to be greatly underestimated; now it appeared that the recoupment would be very much less than anticipated. There is no doubt that the Board found itself under considerable pressure to avoid any delay in disposing of the sites. Much of the land which had been set aside for rehousing under the Metropolitan Street Improvements Act of 1872 still lay vacant, and a campaign had been mounted, particularly by Waterlow, to force the Board to release it for whatever price it could get.[67] The Board was pressed by the Home Office to accept the Peabody offer; indeed Cross said in parliament that 'every power which he possessed would be used to secure the consent of the Board to the proposal'.[68]

The Peabody Trust had little interest in the Whitechapel site, which from its point of view was remote, but it admitted to obtaining the five choicest sites of the Board on relatively favourable terms. Key decisions about the size and arrangement of accommodation in the new buildings passed from the Board to the Trust. Moreover, a condition attached to the Trust's offer was that it should obtain a government loan of £300 000 at the favourable rate of 3.5 per cent per annum.[69] The Trustees themselves, however, maintained that their offer was made solely because 'the areas cleared... were likely to remain useless, the conditions on which the areas are disposable being... such as ordinary builders are not willing to comply with... Even with the advantage of a loan on easy terms, and of a comparatively cheap purchase, they do not calculate on more than a bare avoidance of loss.'[70]

In August 1879 the Board submitted a statement to the Home Office on the workings of the 1875 Act.[71] The Board had 'seen from the beginning that the Act could only be carried out at great expense, but it was not prepared to find that the cost would be so enormous as the figures now before it would show'. The Peabody Trust money would total £91 305 and there was an estimated further recoupment of £81 400 from surplus land which could be sold for commercial purposes. The cost to the Board in compensation and works was, however, £734 766, leaving a net loss to the rates of £562 061. 'The Board is now engaged in carrying out eight other schemes... and the total loss on the fourteen schemes... is estimated to be £1 076 470. Numerous other representations are already before the Board.'

These figures were designed to place the Board's case against the Act in the most favourable light. It was later argued that improved rateable values might be capitalized at £66 400 and that the cost of associated streets should not be charged to the rehousing scheme.[72] The circumstances in which the land was sold may have produced a figure rather below its real value, and 1879 was in any case a year of industrial depression in which dwellings company lettings were more difficult and several major auction sales of land and buildings in London failed to reach their reserve price.[73] However, once allowance has been made for all these factors, the workings of the Act had revealed a financial gap between the purchase and selling price of land, the size of which could not be gainsaid. The loss on the six schemes in the Board's estimate was £36 433 an acre, or rather more than £58 for every person displaced.

The Board had not committed itself to much more than half of the £2m capital expenditure originally proposed by Waterlow, but according to that programme such a sum should have removed half the unfit houses in London. In fact it was obvious that only a beginning had been made, and the Board, frightened by the financial implications of large numbers of further representations, was determined to make a stand. Moreover, although the Board had all the distasteful duties of compensating landlords and displacing tenants, it was now more than ever dissociated from the tangible assets arising with rebuilding.

The Board's solution was put plainly in the August submission. Its architect had put the commercial value of the six sites at £496 694.[74] This would still leave a sizeable loss on the purchase price, and so an attack should be mounted on compensation payments. The Board had 'generally been required to pay for the worst class of property as much as if there was no sanitary reason for its destruction'. The arbitrators should be instructed to consider whether it was 'possible by structural or other alterations to make the buildings fit for human habitation, and if so at what cost', and the amount 'should be deducted from the value of the property in its present state'. If the property could not be so altered the arbitrator should be directed 'to award compensation on the basis of the value of the site of the cleared buildings adding thereto the value only of the materials on the ground'.

The main thrust of the submission was, however, that it was mistaken policy 'to insist that no matter what the value of land in any particular locality may be . . . it must for all time be set apart for labourers' dwellings'. Moreover, 'the new dwellings cannot be provided until the old ones have been destroyed, and in the meantime the poor people leave and are scattered; and it is highly improbable that any considerable number will return'. Thus the new houses will 'be occupied by others who have no particular claim to the locality and might just as well have been accommodated in other parts of London'. The Board therefore proposed that it should have the power 'to dispose of the cleared ground for commercial purposes and to provide ground for rehousing the dispossessed families in other parts of the metropolis'.

Within a space of only four years it had become apparent that Cross's Act had failed, and that any prospect of removing the slum from London in the immediate future had to be abandoned. This could only come about if there were quite fundamental shifts in attitudes to property compensation and public expenditure. The idea that the dwellings companies could achieve it with only minor departures from a market economy was now discounted. Instead, there was the prospect of a long, hard slog, and it remained to be seen what could be salvaged from the wreckage in this context of reduced expectations.

Notes

1   Smith, P. 1967. *Disraelian conservatism and social reform*, p. 200. London: Routledge & Kegan Paul.
2   *Hansard*, vol. 222, HC Deb., 3s., 8 February 1875, col. 99.
3   Whitechapel Vestry 1873. Medical Officer of Health Report.
4   Cross, R. 1882. Homes of the London poor. *Nineteenth Century*, vol. XII, p. 231.
5   Allan, C. M. 1965. The genesis of British urban redevelopment with special reference to Glasgow. *Economic History Review*, vol. XVIII, p. 604.
6   Tarn, J. N. 1969. Housing in Liverpool and Glasgow: the growth of civic responsibility. *Town Planning Review*, vol. 39, p. 329.
7   Charity Organisation Society (COS) 1873. Report of the Dwellings Committee, p. 9.
8   Allan, C. M., op. cit., p. 598.
9   Smith, P. J. 1980. Planning as environmental improvement: slum clearance in Victorian Edinburgh. In *The rise of modern urban planning 1800–1914*, A. Sutcliffe (ed.), p. 113. London: Mansell.
10 Booth, C. 1902. *The life and labour of the people in London*, final volume, p. 211. London: Macmillan.
11 Simon, Sir J. 1897. *English Sanitary Institutions*, 2nd ed, p. 441. London: Murray.
12 Marshall, I. H. 1967. *Social policy in the twentieth century*, 2nd ed, p. 11. London: Hutchinson.
13 Webb, S. and Webb, B. 1929. *English poor law history*, vol. II, pp. 435–60. London: Cass.
14 Simon, Sir J., op. cit., pp. 450, 454, 442.
15 Cross, R. 1884. Homes of the poor. *Nineteenth Century*, vol. XV, p. 157.
16 Hill, O. 1875. *Homes of the London poor*, p. 3. London: Macmillan.
17 COS, op. cit., p. 11.
18 Smith, P. J., op. cit., p. 102.
19 Dyos, H. J. 1982. The objects of street improvement in Regency and early Victorian London. In *Exploring the urban past*, D. Cannadine and D. Reeder (eds), p. 85. Cambridge: Cambridge University Press.
20 Dyos, H. J., op. cit., p. 84.

21 Beames, T. 1852. *The rookeries of London*. Reprint 1970, p. 40. London: Cass.
22 Dyos, H. J. 1982. Railways and housing in Victorian London. In *Exploring the urban past*, D. Cannadine and D. Reeder (eds), p. 105. Cambridge: Cambridge University Press.
23 *The Times*, 2 March 1861, quoted in Rubenstein, D. 1974. *Victorian homes*, pp. 8–9. Newton Abbott: David & Charles.
24 COS op. cit., pp. 7–8.
25 *Hansard*, vol. 223, HC Deb., 3s., 18 March 1875, col. 743.
26 COS 1873, op. cit., p. 3.
27 PRO HO 32 144/1–123, pp. 361–2.
28 COS 1873, op. cit., p. 12.
29 Simon, Sir J. 1887. *Public health reports*, vol. 1, p. 61. London: Sanitary Institute of Great Britain.
30 St Giles Vestry 1858, 1862–3, 1870. Medical Officer of Health Reports.
31 Cross, R. 1882, op. cit., p. 231.
32 Dyos, H. J. 1982. The slums of Victorian London. In *Exploring the urban past*, D. Cannadine and D. Reeder (eds), p. 129. Cambridge: Cambridge University Press.
33 Beames, T. op. cit., p. 26.
34 Engels, F. 1844. The condition of the working class in England. Reprinted in K. Marx and F. Engels, *Collected Works*, vol. IV, pp. 332–3 (1975). London: Lawrence & Wishart.
35 Beames, T., op. cit.; Godwin G. 1854. *London shadows*. London: Routledge; Hollingshead, J. 1861. *Ragged London in 1861*. London: Smith & Elder; Mayhew, H. 1861–2. *London labour and the London poor*. Reprint 1967. London: Cass.
36 Hollingshead, J., op. cit., Preface.
37 Beames, T., op. cit., p. 208.
38 Beames, T., op. cit., p. 44.
39 Beames, T., op. cit., p. 68.
40 Godwin, G., op. cit., p. 10.
41 Beames, T., op. cit., p. 5.
42 Godwin, G., op. cit., p. 75.
43 Bosanquet, C. 1868. *London: some account of its growth, charitable agencies and wants*, pp. 50–1. London: Bell.
44 Hollingshead, J., op. cit., p. 168.
45 *Hansard*, vol. 222, HC Deb., 3s., 15 February 1875, col. 336; ibid., vol. 223, HC Deb., 3s., 18 March 1875, col. 36.
46 Artisans' and Labourers' Dwellings Improvement Act 1875, Section 19.
47 *Hansard*, vol. 222, HC Deb., 3s., 8 February 1875, col. 113; ibid., 15 February 1875, cols 336, 349.
48 COS, op. cit., p. 16.
49 PRO HO 32 144/1–123, pp. 361–2.
50 *Hansard*, vol. 222, HC Deb., 3s., 8 February 1875, col. 102.
51 Hill, O., op. cit., pp. 188, 194.
52 *Hansard*, vol. 224, HL Deb., 3s., 11 May 1875, col. 457.
53 *Hansard*, vol. 222, HC Deb., 3s., 15 February 1875, col. 372.
54 COS op. cit., p. 8.
55 *Hansard*, vol. 222, HC Deb., 3s., 8 February 1875, col. 113; ibid., 15 February 1875, col. 345; ibid., vol. 224, HL Deb., 11 May 1875, cols 457–8.
56 *Hansard*, vol. 222, HC Deb., 3s., 15 February 1875, cols 343, 373, 381.
57 MBW 2411/7 Octavo Printed Reports no. 727 (Holborn), no. 732 (Whitechapel); MBW Minutes WGP 1 Nov. 1875 (39).
58 PRO HO 45 10198/13 31375, p. 5.
59 Estimate of £51 000 in MBW Minutes WGP 8 Nov. 1875 (27), plus £8633 for supplementary acquisition of 25 houses.
60 MBW 2411/7 Report no. 786.
61 Select Committee on Artisans' and Labourers' Dwellings, *PP* VII, 1882, 534; MBW 2411/7 Report no. 32.
62 MBW Minutes WGP 29 Oct. 1877 (39).
63 ibid., 4 Nov. 1878 (79–80).
64 ibid., 24 Feb. 1879 (60).
65 PRO HO 45 10198/B 313755, pp. 7–10.
66 Select Committee, op. cit., *PP* VII, 1881, 5223; MBW Minutes WGP 24 Feb. 1879 (60).

67   *The Builder*, 26 April 1879, pp. 459–61.
68   *Hansard*, vol. 247, HC Deb., 3s., 3 July 1879, col. 1287.
69   Select Committee, op. cit., *PP* VII, 1881, 4367–80.
70   MBW Minutes WGP 30 June 1879 (54).
71   MBW 2411/7 Report no. 915.
72   Select Committee, op. cit., *PP* VII, 1881, 4367–80.
73   *The Builder*, 1879, pp. 325, 723, 783.
74   *The Builder*, 28 June 1879, p. 174.

3 Slums and administrative responsibilities

With the failure of Cross's Act high hopes and spectacular policy initiatives faded from the scene. National governments, unwilling to mount any decisive attack on the financial constraints which lay at the heart of the problem, thrust responsibility for the slum firmly back upon the local authorities. The emphasis was to be on effective administration coming to grips with the slum at the detailed local level in an inevitably lengthy struggle. The rôle of the state would be simply to set the framework of enabling legislation and to see that the local authorities performed their duties. As far as London was concerned, this was to be of considerable significance, for it contributed to the pressures on Parliament to bring the Board of Works to an end and to install a directly elected government for the whole of London. It cannot be denied, of course, that the advent of the Council brought with it a resurgence of expectations, but these were mainly centred on the vigorous nature of its approach and on its determination to fulfill its responsibilities.

Retrenchment by the Board

The Board's bombshell of August 1879 had been unfortunately timed, for it came too late to affect the provisions of an Amendment Act which was already in preparation. However, the Charity Organisation Society found circumstances sufficiently altered to commission a new Dwellings Committee report in 1880, and eventually a thorough review was undertaken by the Parliamentary Select Committee appointed in May 1881. From then on events follow a famous sequence. The Select Committee reported in 1882, and most of its recommendations were incorporated in the Artisans' Dwellings Act of that year. In 1883 Mearns launched his *Bitter Cry* and the periodicals filled with housing articles while the public mind, according to Octavia Hill, was 'in a state of wild excitement'.[1] In February 1884 Lord Salisbury moved the motion which led to the appointment of the Royal Commission on Working Class Housing.

Among the policy developments suggested in this massive outburst of writing, the Board's own preferred solution was put to the Select Committee by G. Richardson. He argued that the rebuilding obligation should be removed altogether, and that the legislature should provide only for the destruction of slums. The land would remain with the owner (and thus not require compensation), but he would not have the right to rebuild except in a sanitary manner: 'Rather than leave a lot of property useless, the owners of property will combine together to get possession of it in a way beneficial to the public generally.' In the meantime, the persons who had been 'in the habit of tenanting these slums' would be 'compelled to put themselves into better houses'. He conceded that there was 'a very small class of labourers' who needed to live near their work, but argued: 'If you use the rates for the purpose of building houses for people near their work, it resolves itself into a rate in aid of wages.'[2]

Enforcement of sanitary legislation in the centre could be coupled with a renewed emphasis on suburban housing provided by the market. This was the line taken by Alfred Marshall and by the minority report of the new COS dwellings committee.[3] It concluded that Cross's Act had proved 'so serious a burden on the rates' that action had been properly suspended. The core of the problem was that 'the average wages of the labouring poor, or even of a better class of artisan', could not bear 'the expense of clearing away costly property and the reasonable ground rents of central sites in the metropolis, and likewise contribute to a fair dividend on the outlay of an improved structure'.

For this reason the output of the dwellings companies continued to be 'as in 1873 relatively insignificant and inadequate in proportion to the needs of the population', nor was there 'any reason to expect a large development in this direction in the future'. Cross's Act might continue to be of some use 'if converted into a Town Improvement Act' but 'reliance must in the main be placed upon the strict carrying out of the sanitary and public health acts and of the (Torrens) Act of 1868'. 'The choice is between attempting to meet the encroachments of commerce by setting aside oases of artisans dwellings . . . or taking a course which accords with what may be called the natural development of large towns, and easing and promoting in every way the gradual dispersion of the people.'

The notion that there was some untapped rent-paying ability which might still be squeezed out of the poor for their improvement was now less generally accepted. Many commentaries, however, started from the position that the application of Cross's Act had brought society 'perilously near to state inter-ference with rents'. Brodrick said, 'Artisans see plainly that ratepayers are already heavy losers by these improvements; it is perhaps natural they should wish then to go a step further . . . by letting improved dwellings below the market value.'[4] Such argument was mentioned only to be forcibly rejected in favour of some other remedy. Brodrick's own preference was an attack on compensation payments: 'The true solution [lay] in the imposition of a heavy penalty on landlords who neglect their duties by denuding them of compensa-tion, except for the fee simple of the ground.'[5]

The most famous political case against the compensation payments was presented by Joseph Chamberlain in 1883.[6] He described Cross's Act as 'by far the most radical and comprehensive scheme of reform that has yet been suggested'. It was, however, 'tainted and paralysed by the incurable timidity with which Parliament, largely recruited from men of great possessions, is accustomed to deal with the sacred rights of property . . . The owners of this class of property, whose greed and neglect have rendered interference necessary, have in every case obtained from the public, under the guise of compensation, amounts altogether and demonstrably in excess of the market value of their property.' Such compensation put 'a premium on evil practices'. It was, he concluded, 'simply a question between the rights of property and the rights of the community [and] the expense of making towns habitable for the toilers who dwell in them must be thrown on the land which their toil makes valuable, and without any effort on the part of its owners.'

Another possibility was a reduction in standards of accommodation. This naturally brought the views of Octavia Hill once more to the fore. She advised a strong application of 'common sense' to the problem, and a dismissal of 'heroic remedies': 'No one can expect to immediately transfer families from homes such as those lately described in newspapers into ideal homes . . . supposing we could so arrange all outward things as to rehouse them, the people themselves are not fit to be so moved, and can only gradually become so.' She advocated 'one large room separable into compartments by curtains or screens'.[7] This was contrasted with the Peabody tenements which one report described as 'elaborate', the result of 'sumptuous' expenditure and devoted to 'the aristocracy of the working classes'.[8] Supporters of the minimum standards approach continued to maintain that there was 'no financial difficulty in providing new dwellings for the very poor on the sites of their old rookeries'.[9] Certainly, Miss Hill believed: 'If blocks such as I describe were multiplied, if the existing laws of demolition were put in force, if sanitary inspection were cheaper, the present difficulties would be to a large extent overcome.'[10]

The official response, perhaps not surprisingly, was to try and adopt something from all these proposals without fully supporting any of them. Indeed, in an effort to reduce public expectations, much emphasis was placed on the lack of any simple or immediately effective solution. As Lord Salisbury put it, there was no 'heroic plan', and 'the best thing to be done' was to 'attack the evil on as many sides as possible'.[11] This did not mean less government. Cross's Act would be modified, and at the same time Torrens's Act, and other sanitary legislation, would be brought back into prominence. The Board of Works would get something in return for its complaints, but the burden of implementing the Housing Acts would not be taken away. It was politically impossible for government to countenance inaction, and impossible to countenance stricter sanitary enforcement without central area rehousing.

So resulted a long battle between central government and the Board. With the strength of its position yet untested, the latter felt able to sweep aside the timid modifications incorporated in the Amendment Act of 1879. In August 1880 it decided that no clearance should take place before a more comprehensive reform. The report of the 1882 Select Committee was thus to prove a considerable disappointment, for in general it reaffirmed belief in clearance and in existing legislation. Although the expense of schemes under Cross's Act was very great, the loss 'as compared with the loss upon an ordinary street improvement would be only about seven per cent more', and against this had to be set the addition to rateable values and above all the improved health and morals of the people.[12]

The Committee sought to allay criticism, however, through a variety of detailed proposals. In provincial towns the rebuilding obligation could be removed, and in London it could, with the approval of the confirming authority, vary from half to two-thirds of the numbers displaced. As for the dwellings themselves, 'in many cases it would be well to have a larger number of single rooms'. The compensation clauses should be revised, and houses dilapidated beyond economic repair should be compensated at cost of land and materials only. Torrens's legislation needed to be put into greater effect, and the operations of the vestries should not be confined to individual houses but 'extended to several houses in a street court or alley'. All these proposals, with the notable exception of the 'land and materials' compensation provision, were brought into effect through the 1882 Act.

This was sufficiently definitive and encouraging to require some response and brought about the immediate approval of four new schemes. However, the Board moved with great caution, and its concern with the size and convenience of the sites is obvious. All four were small compact blocks well removed from central London at Windmill Row Lambeth, Tench Street, St George-in-the-East, Brook Street, Limehouse, and Trafalgar Road, Greenwich. The single scheme with which Cross's Act had begun in Whitechapel involved more persons and much greater cost than all these projects combined. On sites chosen for their low property value, the Board was unable to make use of the remittance of the rehousing obligation. Another corollary was low demand for the land, which was to raise future difficulties.

The more immediate effects of the implementation of these schemes, however, lay not in any damaging new revelations but in the confirmation that action under Cross's Act would always be expensive. The Board's response was a deepened recalcitrance. In July 1883 the Home Secretary informed the Board that he thought it right to call the attention of the vestries to the 'special expediency' of putting the Housing Acts of 1875 and 1882 into 'early operation'.[13] Not surprisingly, this led to a renewed flow of representations, but the Board found reasons to reject them all, and stressed the need to obtain further experience of the workings of the schemes already in hand.

These actions were taking place against a background of mounting public concern over the housing question in London, which culminated in the appointment of the Royal Commission. Unlike the 1882 Committee this was not directed specifically at the problems of slum clearance and indeed aimed to move away from the view that housing problems were 'purely a sanitary question in its most restrictive sense'.[14] Instead, the most important fact they had to face was that 'though there was a great improvement . . . in the condition of the houses of the poor compared to that of thirty years ago, yet the evils of over crowding, especially in London, were still a public scandal, and were becoming in certain localities more serious'. From this perspective action under Cross's Act could hardly appear in a favourable light. Indeed, the Report stated that although demolition was undertaken in the interests of public health and welfare, 'a good deal of hardship is caused by this class of displacement', instancing increased rents and overcrowding in districts such as Spitalfields, the Mint, and St Lukes.[15]

In relation to the clearance programme, however, the Commissions Report was deeply contradictory. Although overcrowding was seen to be the result of demolition, particularly for commercial purposes, this link was in turn produced by the fact that 'an enormous proportion of the dwellers in the overcrowded quarters' were 'necessarily compelled to live close to their work'.[16] From this point of view the analysis was a very traditional one, and it brought public opinion firmly back to the train of reasoning from which Cross's Act had first emerged. It remained the case that the housing problem had essentially to be solved in the central area, and suburban policies were still confined to renewed pressure for workmen's trains.

The Commission therefore never doubted the need for Cross's Act, merely following the Select Committee in its emphasis on a more broadly based approach. To this it added the call for all parties to react more positively to their responsibilities. For the Majority Report the main problem was 'that there has been a failure in administration rather than in legislation, although the latter is no doubt capable of improvement'. The Commission pointed to the need to obtain a reform of local government in London, and in the meantime for the Home Secretary to intervene in disputes between the Board and the vestries over the use of the Artisans' Dwellings Acts. 'Places which are notoriously bad remain so because each authority maintains that the other authority ought to deal with them.'[17]

In the context of the Commission's work, the Board was bound to appear as a body conspicuously failing in its duty. The Home Secretary had, in fact, already anticipated the new proposals, and in February 1884 ordered a local inquiry into the correctness of a representation for Hughes Fields, Greenwich. The report compromised, upholding the representation, but considerably reducing the size of the scheme. Nonetheless, the Board was forced to admit that a large area could only be dealt with effectively by the application of Cross's Act, and it approved the scheme in October 1884. At the same time it adopted a much smaller project for Tabard Street, Newington, which it correctly saw could be undertaken at relatively low cost. This was to be the last scheme which it would voluntarily undertake.

Further local inquiries initiated by the Home Office forced the adoption of schemes for Cable Street, Limehouse, and Shelton Street, St Giles, in 1886, after both had been rejected by the Board. The Cable Street decision was arrived at after 'some division of opinion in the Home Office' but the Shelton Street case was more clear cut. The District Board had already taken action in 1887 under Torrens Acts and 'the houses had already relapsed into an insanitary condition'.[18] The more powerful provisions of Cross's Act seemed to be called for, and the scheme eventually approved was the first to be undertaken in central London since 1876. Even so, the Board's scheme was inadequate, taking a minimum of property and providing a site which proved impracticable for rebuilding. In accepting it, Cross wrote, 'This is by no means so good a scheme as ought to have been presented, but the area is so unhealthy that it would not be right in my opinion to reject it.'[19] On that unsatisfactory note the Board's record of clearance came to an end.

Commitment by the Council

The first elections to the London County Council were won by those who had campaigned for its creation.[20] They were grouped as the Progressive Party – Liberals with a strong radical element, the Conservatives being represented by the Moderates. It was a council determined to make its mark, a body with the political will the Royal Commission had demanded. Indeed, the early history of the Council can be seen as an effort to put the Royal Commission's policies into effect. Whereas the Board had acted mainly as a clearance agency, responding to local initiatives, the emphasis now was on leadership by the Council. It would investigate and publicize unsatisfactory housing conditions of all kinds, and stimulate or pressurize landlords and vestries into appropriate reformative action.

A necessary first step in this direction was the appointment, in March 1889, of a medical officer of health to the Council. This would enable direct investigation to be made with much greater authority. Housing subcommittees were set up for the geographical quadrants of London, and they conducted all the detailed business until the end of 1892. Their existence more or less corresponds with the first active phase of the Council's housing policy. The vestries and their medical officers were urged into action, but the Council was not satisfied with such formal channels of investigation. They were to be supplemented with the help of a whole army of irregulars. Individuals, school board visitors, the Mansion House Council, and many local bodies were asked to report, and when in July 1889 a letter was received enclosing 33 copies of Charles Booth's map of East London, it was decided to buy an additional 75 copies.[21] All this meant that, whereas in the days of the Board proposals had come forward in the form of a limited number of official representations, the Council had 146 areas brought to its notice by the end of its first year of office. In only a minority of cases would there be any question of a clearance scheme. Nonetheless, the sites reported in these early years formed an agenda for such clearance during the rest of the century.

Apart from the investigative effort, the other urgent task for the Council was to reinforce its own powers in the field of public health and housing so that it could intervene directly and bring pressure to bear on landlords and vestries. New legislation sought for that purpose included the Public Health (London) Act of 1891 and the London Building Act of 1894. The most immediate concern, however, was that the powers of the Council in the housing field should not be confined to the implementation of Cross's Act but should also include Torrens's legislation. The aim here was not to take the responsibilities for Torrens's legislation out of the hands of the vestries, but to establish clearly the position of the Council as the supreme overlord across the whole housing field.[22] A deputation from the Council met the Secretary of State in December 1889 to press for these and other provisions, and the Government responded with the Housing of the Working Classes Act, 1890.

The chief immediate importance of the Act did, indeed, lie in Part II which dealt with Torrens's legislation. The powers obtained under Part III, which allowed purchase of land for building outside of clearance sites, were seen as important by the Council, but their implementation was bound to be limited in the context of a policy still firmly concentrated on the central areas. There were few changes to Cross's legislation, incorporated as Part I of the Act, although steps were taken to strengthen the compensation clauses. These went some way towards a reassertion of landholder responsibility. Correspondingly, local authorities were obliged under Part II to seek closure orders for houses on the representation of their medical officers. This, it was thought, would force the landlord to repair, and if he did not would cheapen the compulsory acquisition of his property.

It now became possible to make clearance and rehousing schemes under Part II for areas 'too small to be dealt with under Part I'. Although the primary responsibility for these schemes still lay with the vestries, provision was also made for schemes to be carried out by the Council, with a compulsory financial contribution from the vestry. The Council could also join with the vestry, and contribute financially to an agreed scheme, but the overall effect, compared with Cross's Act, was to throw the costs of clearance back on the local authorities. The Council could therefore direct the course of clearance without having to assume the bulk of the costs. It was envisaged that the contributions of the Council would be modelled on those of the smaller street improvements, varying from one quarter to one half of the costs according to the financial circumstances of the local authority. Above all the advantage of the new Act to the Council lay in the reinforcement of its position *vis-à-vis* the vestries. In a triumphant report the Housing Committee claimed:

> The powers invested in the Council are so large as to practically be co-equal with those of the vestries . . . there is nothing to prevent it at once initiating on a large scale, by means of small schemes, local improvements in all the crowded districts of the metropolis, and at the same time forcing the closure or demolition of every house found insanitary by the medical officer. The position of the medical officer of the Council is no longer equivocal, he is now made master of the situation, and this fact alone is likely to act as a strong incentive to owners of insanitary property to set their houses in order. [23]

The committee pointed out: 'Something over £1m has been contributed since 1856 in aid of local (street) improvements. In 1888 over £57 000 was contributed. A similar yearly contribution in aid of reconstruction schemes under Part II would effect much in a few years to dispose of the worst insanitary areas in the metropolis.' [24] The government were congratulated on 'carrying out so many of the views of the council'. [24]

Meanwhile urgent attention had been given to eight projects which had still been under consideration by the Board. It was decided that most of these should be dealt with under Torrens's legislation. When the vestries refused to co-operate, Home Office arbitration was sought over representations for Shore-ditch and St George the Martyr (Southwark). The reports in November 1890 found, not surprisingly, in favour of the Council. It was recommended that the vestries should undertake schemes under Part II, the Council paying half the cost at Shoreditch and one-third at Southwark. [25] The Part II policy was therefore under way, and the Council had taken important steps towards the implementation of the vigorous but more broadly based approach to housing that the Royal Commission had recommended.

Council members of all persuasions were generally agreed that housing policy should take a pyramidal form, with a broad base of sanitary reform and rehabilitation passing upwards through clearance schemes of various scales. However, there were important disagreements about the size of the whole pyramid and the scope for action at the top. Beachcroft, leader of the Moderates, strongly favoured an approach based on Part II schemes alone, and at one point the Housing Committee recommended that 'pending a full trial of what can be done under Part II no schemes of any magnitude under Part I should be embarked upon'. [26] For others, however, an exemplary scheme under Cross's

Act could provide a flagship for the whole of the Council's operations. In the event the matter was decided very quickly. An official representation for the Boundary Street area was submitted by Bethnal Green in April 1890. In June it was decided that the officials should prepare a scheme for the area, part of Shoreditch was added in September, and by November the Council had committed itself to a scheme under Part I of the new Act involving nearly 15 acres of land and 5719 people.

Boundary Street was to become not only the largest project undertaken by the Council, but also the archetypal clearance scheme of the whole decade, and the manner of its introduction is therefore most important. Here again, the scheme must be thought of as a whole package involving not only arguments over the scale of clearance but also conceptions of reconstruction. On the destructive side, the debate produced arguments of a technical nature, but it also revealed two quite different sources of opposition in principle. The more traditional of these, represented by Beachcroft, maintained that the scheme should be scaled down and undertaken under Part II. Beachcroft argued that to have such a large project in a single parish would torpedo the whole strategy of numerous small-scale clearances.[27] The second source, which was to have an important bearing on the future policy of the Council, was represented by William Saunders, who wished to throw the cost of improvement onto the landlords and saw the Part II strategy as a means to this end. He wrote later:

> In proportion as landlords were alarmed by the improvement of Torrens's Act so they were anxious to continue the application of Cross's Act; landlords, their agents and their friends realized the crisis and threw all their energies into inducing the Council to apply Cross's Act to Bethnal Green. They succeeded, and the ratepayers of London will have the pleasure of buying 15 acres of slumland and selling it again at a net loss of £300 000 . . . The Board of Works had this excuse, that they were without efficient modes of compelling landlords to improve their property. The Housing Committee claim credit (which they deserve) for obtaining such powers . . . and . . . still go on with Cross's Act.[28]

On the technical side it was argued that 'the area could not be dealt with except as a whole', and that owing to its size it should be considered as a metropolitan improvement.[29] Beachcroft had maintained that schemes should not be given Part I status simply because they were large. The legislation, however, envisaged that Cross's Act would continue to be applied, merely shifting the boundary between Torrens and Cross upwards in a rather vague manner. Although the Home Office generally supported the Part II strategy, it was not clear that it would do so whatever the size and condition of the area represented.

Although the Council did not undertake a reconstruction scheme itself until 1892, it is important to realize that considerable steps had already been taken in this direction before the Boundary Street operation was under way. Indeed, because the Council had inherited six uncompleted sites from the Board, the most pressing decisions it had to take involved rebuilding rather than new clearances. Only one of the six sites at Shelton Street had a central position; the others were dockland areas in East London. The Council soon found that the

Shelton Street scheme required modification, and in the meantime none of the other sites was very attractive to potential builders. The Board had already tried to sell Brook Street in 1887 and Tench Street in 1888 with no success. Stewart, in his account of the Council's operations published in 1900, says, 'The sites were not looked upon favourably by persons engaged in erecting working-class dwellings, and in these circumstances the Council resolved to fulfill its statutory obligations by itself erecting the dwellings.'[30]

This step was not, however, simply a pragmatic response to particular difficulties. If it owed much to the practical consideration of the moment, there was also now within the Council a group which favoured in principle the direct building and letting of working-class dwellings. The Council of the London Liberal and Radical Union, meeting at St James Hall in February 1889, had passed a resolution that 'ample compulsory powers should be vested in the County Council for the purpose of enabling them to obtain land in fee simple and erect and hold dwellings for the industrial classes and manage and let the same direct tenants of those classes'.[31] The Council did not begin with any prolonged testing of the market. In June 1889 the Solicitor was asked to report the legal situation. Then, despite considerable divisions, and a noticeable failure to agree the principles on which future rent scales should be based, in July 1889 the Council decided to apply for permission to build dwellings itself at Hughes Fields.[32]

This move proved to be premature. It met considerable political opposition from vestries such as St Lukes and Hammersmith. Without making a decision in principle, the Home Office refused the plans remarking, 'It is not desirable that the Council should themselves undertake an operation of this nature unless the buildings supply accommodation suited to the class for those whose benefit the scheme was intended, in a manner which may serve as a model for other districts.' The Council in their reply agreed that building by them as 'an experiment' should serve as a model for other districts. 'They have also inferred from your letter that . . . it may be wiser that such experiment should be tried in some other locality where more immediately pressing need exists for further accommodation.'[33]

These developments were important pointers towards future policy. Hughes Fields was said to have much vacant working-class accommodation in the neighbourhood and to offer a poor market for Council dwellings. This emphasis on the low demand for housing on suburban sites pointed the Council towards areas nearer the centre, and told heavily against any nascent support there may have been for building in the suburbs under Part III of the Act. It was by no means evident that such schemes would be successful, and in any case they would be very difficult to justify to the Home Office at this stage. The Boundary Street area, however, was in many ways an ideal location for a Council housing project. It lay close enough to the City to ensure sufficient demand, and yet would not involve the very high costs of property acquisition in districts such as Holborn or St Giles.

Important decisions on housing standards were also made in 1889. Here the trend was towards lower-density usage of sites and higher standards of accommodation. It was the one main area in which the Council's policy ran against the trend of the mid-1880s. It was fully in accord, however, with the views of the

Home Office, which emphasized the need for any Council development to serve as a model, and required a height limit of four storeys and minimum room sizes. These conditions were accepted by the Council and further stipulations were added, requiring, for example, a greater distance between buildings.[34]

Stewart, in his account, says that the Council felt it necessary to 'secure the erection of tenement dwellings which would in all respects bear the closest investigation. Any dwellings that the Council approved of should be of the best description.' Moreover, 'sanitary science was advancing by such rapid strides that it was essential that the Council should keep in view what was likely to be the sanitary standard in the future'. He admits that 'these conclusions may appear to be at variance with an intention to rehouse the lowest class of the working population', but justifies this by referring to evidence given to the Royal Commission that the class of persons displaced could not be re-accommodated in rebuilding schemes.[35]

Although the positive reasons which Stewart advances for the adoption of relatively high standards are no doubt correct, it is less certain that the Council took a clear-cut hard-headed decision not to cater for the persons displaced. Indeed, it seems rather more likely that they simply fudged the issue, accepting the reasons for high standards but still clinging to the belief that some of the poor could be accommodated. It has to be remembered that at this stage the Council was mainly thinking about cleared sites and was not much involved in the actual displacement of tenants. The crucial issue of rents was simply postponed. Moreover, there was considerable optimism in some quarters about the effects of Council building. Dodd, writing in a Liberal publication of 1889, thought it useless 'to turn people out of insanitary dwellings until sanitary ones are provided'. These buildings should not be 'diverted to the use of those who are richer'. This did not in his view mean a reduction in standards, however, for 'lifts will be needed in the block buildings and buildings must be made more home-like'. This could be done because 'the middle man . . . and the profit making company will be avoided. The County Council will become a landlord . . . giving a good dwelling at a fair rent.'[36]

The Boundary Street decision was, therefore, not fundamentally a matter of technicalities. It has to be interpreted as an act of faith in the powers of the Council, a desire to set out on a new path and to affirm the Council's leadership. In adopting the scheme the Council was squarely facing its own responsibilities and signalling to the vestries and landlords that there was no excuse for them not to follow, even at some financial cost. There was, moreover, considerable confidence in the Council's ability to carry out an exemplary scheme. Here was the opportunity to mark the distinction between the Council and the Board clearly, and it was no doubt to the advantage of the scheme that it was a new proposal not inherited from the Board, and from a parish which had seen no previous clearance. The potential for erecting municipal dwellings on municipal land was very attractive to many of the more radical members who might otherwise have supported the Saunders camp. All these arguments in favour of the scheme convinced Benn and even Fleming Williams, who nonetheless hoped that 'the time was not far distant when Londoners would insist that the ground owners should contribute something to the cost of the improvements'.[37]

Disillusion, 1891–8

At the end of 1890 the Council could justifiably claim substantial progress on all fronts: at the lower level a mass of cases resulting from the investigative effort were being followed up, the Part II policy was underway, and the Council was setting an example with a major scheme of its own. Once again, however, it required only a few years for this position to crumble, and by the end of 1893 it was in complete disarray. Indeed, as early as April 1892 the north-east subcommittee rejected a scheme for Great Pearl Street, Whitechapel, 'having regard to the present state of the law in respect to compensation paid to owners of property'. They could not recommend the Council to accept schemes 'except under the direct compulsion of the Home Secretary'.[38] Yet this was the same subcommittee that had enthusiastically promoted the Boundary Street scheme such a short time before.

Boundary Street, indeed, overshadowed all other developments in this period. It did not cost more than foreseen – indeed somewhat less – but there was a long period between the excitement of the adoption phase and an eventual political return in the form of visible assets. Meanwhile, the more disagreeable stages of displacing tenants and compensating landlords had to be faced. Almost immediately after the commitment to Boundary Street was made the Council began to show signs of nervous withdrawal and to become more hesitant in its clearance policies. By the end of 1892 an embarrassing backlog of possible schemes had developed, and the regional subcommittees which had promoted local investigations were wound up.

The vestries, too, were enthusiastic for clearance only when the costs were borne by others. It soon became apparent that Part II schemes allowed abundant scope for prolonged haggling between the authorities concerned. At Shoreditch a scheme for Moira Place and Plumbers Place was finally agreed in 1893. At St George the Martyr, Southwark, the vestry began by maintaining that the areas for which they had demanded a Part I scheme could now be dealt with by closure orders alone. They suddenly found that there was 'abundant accommodation in the neighbourhood for the working classes'.[39] On the insistence of the Council they nonetheless agreed a scheme for Gun and Green Street in 1891, but action on the remaining area around Falcon Court was delayed until 1895. Due to the pressure of the Council, further schemes resulted at Brookes Market (1891), Mill Lane (1892), Ann Street (1893), Norfolk Square (1892), London Terrace (1892), and Queen Catherine Court (1893). The Council undertook the first three of these projects and the vestries the remainder, the costs in all cases being equally shared. All these negotiations with more or less recalcitrant authorities began in 1891 and occupied much time, despite the fact that the scale of clearance and net expenditure were both very low. Even after a scheme was agreed disputes often arose during the course of implementation.

The key struggle developed around representations made under Part I for schemes in the Somers Town district of St Pancras. These formed, by their sheer size, the most important test of the possibility of a clearance programme based on Part II schemes since Boundary Street itself. Significantly, the initiative was taken by the vestry, which rejected the Council's decision of May 1891 that the areas should be dealt with under Part II and applied for Home Office arbitration.

In the event the Home Office stood by the Council and on their own admission leaned as far towards its view as possible.[40] The outlying parts at Prospect Terrace and Brantome Place were assigned to the vestry. The central area, lying between Euston and Kings Cross stations, was divided in two, the western portion around Churchway to be dealt with by the Council under Part I and the rest by the vestry. The outcome was inevitably unsatisfactory to both parties. The vestry had been given much the larger share of the burden, but it lay with the Council as the higher authority to set an example and begin the actual work of clearance.

When the Churchway scheme came before the Council in October 1893 it was rejected on the following grounds: 'Parliament has not . . . by an improvement rate, taxation of ground values, or rate other than that falling on the occupier . . . provided the Council with sufficient resources to carry out the schemes.' It was resolved that 'the owners of the ground values of the area, viz. the Trustees of Lord Somers and Lord Southamptons estates and Lady Henry Somerset, the present life tenant, be informed of the representation and that they be urged to take the necessary steps to put an end to the present condition of things'.[41] Here the Council's stubborness was merely part of a wider campaign over land values which came to a head in 1893 and brought municipal improvement in the capital to a temporary halt. Another major battleground was the long-running dispute over betterment in the projected Kingsway Street improvement between Holborn and the Strand.

Developments on the reconstruction side appeared scarcely more satisfactory. After negotiations with dwellings companies in April 1892 the Council itself applied to build at Brook Street and this was shortly followed by similar applications for Cable Street, Hughes Fields, and Shelton Street.[42] This time all the applications were accepted by the Home Office, since it was felt that the overriding need was to begin some reconstruction on sites which had been vacant for so long. By April 1893 the Council was ready to announce its first building at Boundary Street, but before this could be done it was necessary at last to tackle the issue of rents. In their policy document of March 1893 the Housing Committee claimed three main objectives: '1) The provision of accommodation for the poorest class in buildings of sound construction with good sanitation and some attractiveness of exterior, 2) the fixing of such rents as are fair having regard to the rents paid in the neighbourhood . . . 3) the obtaining of a fair percentage upon outlay.'[43]

There were obvious difficulties in reconciling such objectives. Although the Council could borrow at lower rates of interest than outside bodies, it was handicapped 'by the necessity for observing proper regulations of procedure', and by the need to see that its dwellings were 'from a sanitary point of view a model for others to imitate'. The committee considered whether rents should be charged at less than full market value, but rejected this possibility: 'The benefit the tenants will enjoy as a result of the Council's action will be proper sanitation and increased comfort rather than a reduction in rent.' Instead, neighbourhood rent levels were to be made a deciding factor in determining the nature of construction, and rents were to provide a net return of not less than 3 per cent.

This 'three per cent resolution', with some modifications, controlled the development of Council building throughout the rest of our period. Five storeys instead of four became the general rule in Council building, and the Home Office was forced to acquiesce. At the same time the prospect of rehousing the displaced now seemed finally to disappear. Even before its rents policy was decided, however, the Council had encountered major setbacks to its rehousing plans at Boundary Street. Although the clearance programme was phased, attempts to move tenants into the suburbs or into existing model dwellings failed. Development of an initial rehousing site at Goldsmith's Square ran into insuperable difficulties, and in the end tenants had once again to find their own way into alternative housing, although this was done on a more controlled basis than before.

The decisive failure of the Council's rather feeble attempts to rehouse the displaced, coupled with the difficulties over the cost of schemes, necessarily strengthened the two strands of opposition which the Boundary Street debate had revealed. By 1893 Beachcroft 'did not believe that the ratepayers would allow the Council to spend much more money on clearing insanitary areas. The enormous cost made it hopeless to endeavour to rebuild London on those lines'.[44] Even his enthusiasm for Part II schemes was diminished because their cost still had to be borne by the rates. On the Progressive side Pickersgill intervened in the rent debate to ask 'Why should the Committee draw so wide a line . . . between the duty of the council in clearing away insanitary areas and the duty . . . in respect of rehousing those displaced by such clearances?' The policy was 'in conflict with the public professions of many of them'. Frequently the clearances did more harm than good, aggravating overcrowding in the surrounding districts, and there were considerations 'which were paramount even to questions of deficit'.[45] This was not a popular view, however, and instead the more radical wing of the Progressives was already veering towards a suburban solution, encouraged by William Thompson's experiment with Part III housing at Richmond begun in 1892. William Harrison, for example, was asking 'what proportion of the working class could properly be housed outside central London on cheap lands', and demanding a Council building programme 'ungoverned by the accident of the mere possession of cleared spots'.[46]

The 1895 elections produced a split result, although the Progressives continued to rule through their aldermanic majority. Increased Moderate influence was felt in renewed attempts to let clearance land to dwellings companies and in the timid nature of proposals put forward in 1896 to amend compensation procedures.[47] Nothing came of either of these ventures. The elections did serve, however, to unblock the freeze on improvements. At Churchway negotiations with Lord Southampton failed, but an agreement with Lady Somerset was incorporated in the scheme adopted by the Council in October 1895. It was not until March 1899 that it was learnt that the Court of Chancery would not sanction the agreement, so the Council had to proceed with the scheme alone.[48] The Kingsway project was also brought back into consideration and, again in October 1895, the Council agreed on a Part I scheme for slums adjoining the proposed route in the Clare Market area. The bulk of the rehousing for this scheme was to take place on the Millbank prison site, whereas most of the Clare Market land would be released for commercial development.[49]

Clare Market was thus a very special case and although its adoption, coinciding with that of Churchway, gave a new boost to clearance, it by no means heralded any real change of attitudes. More significant was the stalemate between the Council and the vestries over Part II schemes. Although there were new schemes in 1897 for John's Court, Limehouse, and Fulford Street, Rotherhithe, both these districts had originally been represented in 1891 and 1892 and there was a noticeable lack of progress in formulating new projects. Although Beachcroft had been right in his prediction that the Boundary Street scheme would undermine the strategy of numerous small-scale clearances, his opponents had been equally right to argue that a Council that was reluctant to undertake clearance itself would be in a weaker position to force such policies on the vestries. The Council had to lead through some schemes of its own, and in the absence of an agreed suburban solution there remained only the policies envisaged by Cross's legislation.

In the event the stalemate was broken by changing relationships between the Council and the government. Already in 1895 'tenification' had become an election issue as the Council sought to respond to Conservative plans to curb its powers by replacing the vestries with more powerful London boroughs. The whole future of the Council was seen to be at stake, although when reform eventually came in 1899 the creation of the new boroughs involved only a limited weakening of the central authority.[50] It was clear, however, that the Council would never recover the kind of paramountcy over the unreformed vestries that it had so triumphantly proclaimed in its early days. Its advocates needed now to stress the kind of rôle that could only be performed at county level. When the Progressive leader McKinnon Wood wrote his defence of the Council for the election of 1898 he made little mention of action under Part II of the 1890 Act. Instead he emphasized that

the larger and more expensive schemes could not be carried out by local authorities . . . Yet it is to the benefit of the whole metropolis that such plague spots should be swept away and the cost spread over the whole of London is easily borne. Despite what has been accomplished the work is only well begun.[51]

The final fling

The Progressive election victory of 1898 was based on a reassertion of the Council's rôle and authority in the face of adverse pressure. In that sense a political climate reminiscent of its early years was reproduced, together with an expectation of vigorous action, not least in the field of housing. The exact form such action should take was less easily established in view of divisions within the ruling party. By the end of 1898, however, a compromise was worked out within the context of an agreed need to increase housing supply.

This enabled the radicals to grasp their prize in the form of an extended use of Part III of the 1890 Act. These powers, which had hitherto been applied only to rehousing obligations, would now be used to acquire land for general housing purposes. A second prong of attack, however, reflected more traditional concerns, for it involved a return to the strategy embodied in Cross's Act of providing new housing for old on slum clearance sites. In line with changed priorities such sites would be used to rehouse the entire numbers displaced. It was recognized that this would involve an increase in the cost of schemes, and the claim that such costs should be met by London as a whole, and not increase the burden of the poorer districts, formed part of the justification for a return to Part I policies.[52]

In May 1898 the Council's plans under Part III had come under heavy internal fire from its principal officers.[53] They emphasized that Council building should be designed essentially to meet rehousing obligations, and that it should not compete with private enterprise in the provision of additional dwellings. For the medical officer, citing census data on overcrowding, it was 'perfectly obvious that a public authority such as the council cannot itself provide a sufficient quantity of dwellings to make any impression on these figures'. It might indeed have a deterrent effect. The architect emphasized that 'the idea of removing the present inhabitants of an overcrowded area to the suburbs on any considerable scale would be impracticable . . . the experience of many of these persons, their habits and even their means of existence are identified with the place where they now live.'

The valuer, though recognizing the existence of a widespread housing problem, stated: 'It must not be supposed that the mere provision of a greater number of healthy dwellings for the working classes, whether by the Council or by any other means, will of necessity remove the evil . . . healthy dwellings cost more than unhealthy dwellings. Poverty and high rents have as much or more to do with the question than the available amount of house accommodation.' The basic problem arose from the growth of commerce and industry, and resultant movements in population and land values, which were the 'natural results of the working of economic laws'. Because the Council was allowed to write down land values under Part I schemes, 'in those localities where land values are comparatively high, and where there is a demand for dwellings of a superior class, the council can build dwellings satisfactorily'. Here it could reconcile its obligations to building standards and financial rectitude. 'But where land is less valuable (that is to say in the intermediate zone and not on the outskirts) and cheaper dwellings are needed, I have found that it is not possible to comply with all the conditions if the Council itself builds . . . it would be more difficult for the Council to build so as to comply with its financial conditions in the outskirts than it has found it in the intermediate zone.'

Such remarks were meant to recall the difficulties the Council had met from its early days regarding effective demand for buildings on outer sites such as Hughes Fields. What the valuer was, in effect, arguing was that because of its relatively high building standards the Council could not compete with private enterprise except on subsidized land, and this was only available through special deals, as at Millbank, or on slum clearance sites. The official's advice was that on the one hand the Council should not compete with private enterprise and on the other that it could not. As a simple attempt to ward off the new housing policy this was fair enough, but given that the Council was committed to the use of Part III, such advice pulled in contrary directions in terms of the location of effort.

If the justification of Part III was to be the failure of the market, then it might seem that effort ought to be concentrated on those localities where failure was most evident. The Housing Committee was first tempted along this line. Its policy subcommittee recommended in July that it was 'not desirable to erect dwellings outside the county of London'.[54] In its first report in November, the committee accepted that on the outskirts the Council should 'not interfere with work which is now being done fairly efficiently . . . the position nearer the centre is, however, different, the demand for houses is very great, and there are at present very few persons undertaking the supply'. The Council should purchase under Part III 'any plots of land in the county, when it can do so in accordance with the financial regulations of the Council, and proceed to build thereon'.[55]

The Council began, however, from the basis that 'in building to supply houses when under no statutory obligation to do so, there should be no charge upon the county rate'. The logic of this was that the Council should build in the localities where private enterprise built, and it was recognized that 'in strong contrast to the cessation of building of workmen's dwellings in the centre of London' there was a 'very remarkable increase in the number of cottage dwellings which have lately been erected in some districts outside the county'. Another corollary was that the Council would have to align its building standards more closely to the market. The committee made the point that the Council had hitherto opposed 'the policy of building houses of an admittedly inferior type', but if it desired 'to house the poorest classes in Central London it will have to approach more nearly to the Liverpool standard of Council building. In the outer circle also cottages will have to be built to a standard more nearly resembling that of the dwellings with which they have to compete.'[56]

It seems clear that there was initially much confusion on these issues, and a wide variety of argument and expectation. David Waterlow, son of Sir Sidney and future Progressive housing chairman, did not share the hopes of many of his colleagues as to cottage dwellings in the suburbs: 'More important than this suburban policy is the provision in the central portions of London of artisans' dwellings on a slightly lower standard than the present but not so low as the Liverpool standard.'[57] George Haw wrote: 'Not only must cheaper block dwellings be built in inner London for many who must live near their work, but cottages are wanted in the outskirts and even beyond London.'[58] Meanwhile Wallace Bruce, four times chairman of the Housing Committee, believed that there were 'hundreds of acres in East and North-East London covered with small two-storeyed houses which should be rebuilt with tenement building of five storeys'.[59]

Although motions that the Council build 'within and without the County' were again refused, the eventual resolution adopted by the Council contained no geographical reference. The search for sites which then began operated over a very wide geographical area and many conditions of site. In March the Council agreed to accept an offer of land at Edmonton, made by Sir Samuel Montagu. This raised the question of the Council's right to acquire land outside the County. Legal opinion being negative, the necessary authority was not obtained until the Housing of the Working Classes Act, 1900. In the meantime, suburban sites within the county were under prospect. One at Totterdown Fields, Tooting, attracted early attention, but negotiations dragged on, such that its purchase could not be approved by the Council until January 1900.

Much more serious problems confronted the search for sites within the built-up area. Although many parts were 'at present occupied by only small houses with gardens, it would be difficult for the Council to accommodate more persons on them than are at present crowded into these houses'.[60] Even where considerably more tenants could be rehoused than displaced, as at George Gardens, Bethnal Green Road, the valuer reported that 'this was a typical example of sites in such parishes as Bethnal Green where the Council could not give the market price for the land and erect dwellings upon it to comply with the standing orders'.[61] There were exceptions of course, where the flexibility in choice of sites offered by Part III was an advantage, and afterwards two purchases were made in Holloway, one on a site formerly occupied by the Royal Caledonian Asylum. In general, however, it soon became clear that Part III building would be forced into the same suburban pattern as private building, and the options polarized into central area building on slum sites or suburban building on green fields.

Meanwhile, Part I sites were readily available through the already established machinery of representation. The first positive results of the Council's new policy thus lay in the adoption of Part I schemes. In October 1899 it committed itself to the largest programme of slum clearance undertaken in a single year in London since 1876. Together with a special scheme in St Marylebone, under-taken in conjunction with Lord Portman, these schemes had an estimated net cost of £506 900 and were to involve the displacement of 4347 people. The principal sites were at Webber Row (Southwark), Aylesbury Court and Union Buildings (Clerkenwell and Holborn), and Garden Row (St Lukes, Clerken-well). The last two schemes, in particular, involved the acquisition of a good deal of commercial property and meant tackling long-standing clearance problems on sites first represented in 1875 which even the first Council had not felt able to deal with.

This rather heroic commitment passed through the Council with surprisingly little opposition. Beachcroft and Fleming Williams moved an amendment that the Council proceed by means of closure orders. Their views were, however, countered by others like Crooks who said that the Council 'should not keep these people in filth any longer', and in the end 'only four hands were held up in favour of the amendment'. The various recommendations were then adopted without any serious opposition.[62] However, many in the Progressive ranks reflected with some consternation that, instead of moving rapidly onto the promised land of Part III, the Council was once more plunging itself into the unprofitable slough of Part I. 'No human power', complained the *Municipal Journal*, 'seems capable of making the Council understand that it is Part III of the Act that London needs above all.'[63]

The parallels with the Boundary Street debate were obvious. Once more the emphasis on the Council's responsibilities carried the day, but again there was a very rapid cooling of enthusiasm immediately afterwards. These schemes underlined the fact that familiar problems of costs and rehousing had in no way changed. The terms of compensation remained the same; indeed these schemes were more costly than Boundary Street, due largely to their location. The Home Office rubbed salt into the wound by insisting that all commercial property be treated as neighbouring lands, with extra compensation. On the rehousing side some prospect of development had seemed to be opened up by renewed interest in cutting standards and costs, but in the end this too could have only a marginal effect.

Soon after the election the Council had cut its standards in certain buildings designed for rehousing from Clare Market. Liverpool corporation had adopted a policy, to be discussed in Chapter 4, by which Council building was to be specifically designed to rehouse those displaced from slum clearance programmes. Although willing to compromise on standards, however, it was never likely that the London Progressives would follow the Liverpool line. The Council's architects argued consistently against the 'cheap and insubstantial character' of the Liverpool buildings. Above all, although the Council could have reduced costs and rents by lowering standards, there was no prospect of reconstruction at rents which would have allowed some general commitment to rehouse the persons displaced. Indeed, rising interest rates c.1900 made the problem of reconciling standards and financial obligations yet more difficult. The Council joined with other municipalities in seeking an extension of the sinking-fund period, but this produced little response.[64]

Despite the attempts to present it in some novel context, the Council's slum clearance programme thus appeared very much the same mixture as before, open to the same objections. When the Prince of Wales formally opened the Boundary Street estate in March 1900 he gave his seal of approval to a project which was already beginning to seem part of a bygone era. The Tooting scheme, approved in January, involved the acquisition of nearly 39 acres for £44 238, a sum which could not fail to be compared with the £500 000 recently voted for the acquisition of 12¼ acres of central area slum land. The public campaign for Part III policies reached its peak in 1900 in a series of meetings, such as one on Plumpstead Common in September, called by the Woolwich District Trade and Labour Council and supported by Fred Knee, secretary of the Workman's Housing Council.[65] Before the elections of 1901 the Council was able to announce the purchase of 226 acres at White Hart Lane, Tottenham, and 30 acres at Norbury.

The swing away from slum clearance among the Progressives was marked, as it had been in the past, by renewed emphasis on legislative reform. In January 1900 the Housing Committee was asked for proposals 'to compel the owners of insanitary property to bear the whole cost of rehousing all persons displaced from such areas'. Special interest was attached to the joint venture with Lord Portman at Nightingale Street, St Marylebone, since it was the first in which the Council had succeeded in co-operating with the freeholder of an insanitary area with a view to 'carrying out a scheme at the freeholder's cost'.[66]

The Council's eventual 'Freeholder Scheme' proposed: 'The law should declare that it is the duty of the freeholder to see the dwelling houses on his property are fit for human habitation.' Provision should be made for him to enter into full possession of the land (the local authority meeting the compensation costs of the leases) and to carry out a scheme subject to a rehousing obligation. If he refused to do so, he should be regarded as failing in his duty, and the local authorities should be empowered to acquire the land, paying only its market value subject to the obligation to rehouse persons of the working classes.[67]

In February 1901 the Council passed a resolution seeking to amend the 1890 Act so that in the event of 20 or more persons being displaced by the operations of private owners in demolishing dwelling houses, obligations similar to those imposed on railway companies and public authorities should be imposed upon such private owners.[68] It was envisaged that such rehousing obligations might be met by contributing to a Purchase of Site Fund set up by the Council, which would have enabled the Council to supply the requisite housing. Like other legislative proposals of the period this met with no response from central government, whose initiatives the housing field had by now almost completely dried up.

The small scheme at Providence Place, Poplar, adopted in October 1901, represented the Progressive Council's last venture in Part I schemes. In July 1902 it decided to add to the list of proposed amendments to the 1890 Act 'that only one-third of those displaced by a scheme should necessarily be rehoused in the area'.[69] This seems to mark the definitive demise of the Part I section of the 1898 policies. In the meantime the Council set its face firmly against any new proposals. Not unnaturally, its declared policy of 1898 had produced a renewed flow of representations, including notably Devonshire Place and Burne Street, St Marylebone, Digby Street and Brady Street, Bethnal Green, and Warner Street – the 'Italian colony' – in Holborn. All these schemes were warded off. The Boroughs appealed to the Home Office, but this time in vain, official inquiries finding in favour of the Council. By then the Council's view was beginning to be reflected in official reports. That of the Royal Commission on London Traffic (1905) referred to

a typical urban clearance scheme and a typical land purchase and construction scheme in the suburbs – in the one case a great waste of public money and a still crowded population per acre – in the other case no loss of money at all and a population housed in healthful surroundings.[70]

Notes

1   Hill, O. 1883. Common sense and the dwellings of the poor. *Nineteenth Century*, vol. XIV, p. 925.
2   Select Committee on Artisans' and Labourers' Dwellings, *PP* VII, 1881, 5277–86 and 5460–3.
3   Marshall, A. 1884. The housing of the London poor I: where to house them. *Contemporary Review*, vol. XLV, pp. 224–31; COS Dwellings Committee 1881. Report, pp. 137–41.
4   Brodrick, W. 1882. The homes of the poor. *Fortnightly Review*, vol. XXXII, pp. 420–31.
5   ibid., p. 429.
6   Chamberlain, J. 1883. Labourers and artisans dwellings. *Fortnightly Review*, vol. CCIV, pp. 761–76.
7   Hill, O., op. cit., pp. 926–9.
8   Mulhall, M. 1884. The housing of the London poor II: ways and means. *Contemporary Review*, vol. XLV, p. 232.
9   Hoole, E. 1884. The housing of the London poor III: cost of tenements. *Contemporary Review*, vol. XLV, p. 238.
10 Hill, O., op. cit., p. 931.
11 Salisbury, Lord 1883. Labourers' and artisans' dwellings. *National Review*, vol. II, pp. 312–3.
12 Select Committee, op. cit., *PP* VII, 1882, xxi.
13 MBW Minutes WGP 23 Jul. 1883 (54).
14 *Hansard*, vol. 234, HC Deb., 3s., 4 March 1884, col. 1679.
15 Royal Commission on the Housing of the Working Classes, *PP* XXX, 1885, 4, 20.

16 ibid., p. 18.
17 ibid., pp. 36, 34.
18 PRO HO 45 10198/B 31375, pp. 50–2.
19 ibid., p. 54.
20 For the background to the creation of the Council *see* Young, K. and Garside, P. 1982. *Metropolitan London: politics and urban change 1837–1981.* London: Edward Arnold.
21 LCC Minutes HC 29 July 1889 (25).
22 The Council's objectives are set out in LCC Minutes 5 Nov. 1889 (13) and 22 Apr. 1890 (14).
23 LCC HSG/GEN/2/3 no. 13.
24 LCC Minutes 13 Jan. 1891 (8); Minutes HC 13 Oct. 1890 (9).
25 LCC Minutes 22 Apr. 1890 (14); Minutes HC 31 Mar. 1890 (22); ibid., 24 Nov. 1890 (18).
26 LCC HSG/GEN/2/3 no. 13.
27 LCC HSG/GEN/2/3 no. 15.
28 Saunders, W. 1892. *History of the first London County Council 1889–1891.* pp. xlvii–iii. London: National Press Agency.
29 Saunders, W., op. cit., pp. 350–5.
30 LCC 1900. *The housing question in London 1855–1900,* J. Stewart (ed.), p. 47. London: London County Council.
31 St James Hall Resolution no. 3, 14 Feb. 1889, reported in Dodd, J. 1889. *The housing of the working classes.* London: Liberal Publication Department.
32 LCC Minutes HC 17 Jun. 1889 (3); Minutes 2 Jul. 1889 (21).
33 PRO HO 45 10198/B 31375, p. 46.
34 LCC Minutes 3 Dec. 1889 (11).
35 LCC 1900, op. cit., pp. 43–4.
36 Dodd, J., op. cit., p. 17.
37 Saunders, W., op. cit., pp. 356–9.
38 LCC Minutes HC, north-east s/c. 5 April 1892.
39 LCC Minutes HC, south-west s/c. 16 June 1891.
40 PRO HO 10198/B 31375, p. 67.
41 LCC Minutes 31 Oct. 1893 (9).
42 LCC Minutes 5 April 1892 (15), 25 Oct. 1892 (12), 8 Nov. 1892 (16).
43 LCC Minutes 10 Mar. 1893 (9).
44 *London (Municipal Journal),* 15 June 1893, p. 313.
45 ibid., 24 March 1893, p. 107.
46 ibid.
47 LCC Minutes 7 July 1896; LCC 1900, op. cit., pp. 60–2.
48 LCC 1900, op. cit., pp. 217–9; LCC Minutes 19 Feb. 1895 (13).
49 LCC 1900, op. cit., pp. 221–2.
50 Young, K. and P. Garside, op. cit., pp. 64–105.
51 Wood. T. M. 1898. The attack on the council. *Contemporary Review,* vol. LXXIII, p. 212.
52 LCC Minutes 1 Nov. 1898 (25).
53 LCC HSG/GEN/2/2 nos 22–4.
54 LCC Minutes HC 8 July 1898 (2).
55 LCC Minutes 1 Nov. 1898 (25).
56 LCC Minutes 29 Nov. 1898 (8).
57 *Municipal Journal,* 19 Jan. 1899, p. 71.
58 Haw, G. 1900. *No room to live: the plaint of overcrowded London,* p. 155. London.
59 *The Spectator,* 15 Dec. 1898, p. 906.
60 LCC Minutes 21 Feb. 1899 (17).
61 LCC Minutes HC 26 Apr. 1899 (20).
62 *Municipal Journal,* 27 Oct. 1899, p. 1179.
63 ibid., 20 Oct. 1899, p. 1149.
64 The maximum was extended from 60 to 80 years in 1903, but only for the land element.
65 *The Builder,* 1 Sept. 1900, p. 199.
66 LCC Minutes HC 24 Jan. 1900 (10); Minutes 9 Oct. 1900 (15).
67 LCC HSG/GEN/2/3 no. 7.
68 LCC Minutes 20 Mar. 1901 (36).
69 LCC Minutes HC 11 July 1902, Report of s/c. on 1890 Act.
70 Royal Commission on London Traffic. Report, *PP* XXX, 1905, 11.

4 Slums and suburbs

The purpose of this chapter is to review the course of policies and programmes since 1875 and to embody them into some larger logical structure. Clearance schemes need to be disengaged from the rush of events and placed in more general spatial and chronological contexts. Above all, the object will be to examine how the clearance strategy broke down and a rival paradigm came to be established. Attention is first given to the discussion of urban structure and processes in the work of Charles Booth. This was not only the predominant account of its age, but both mirrors and informs the political thought and action discussed in earlier chapters.

Charles Booth's London

Booth's view of Boundary Street was that 'the place deserved destruction. A district of almost solid poverty and low life in which the houses were as broken down and deplorable as the unfortunate inhabitants, it seemed to offer a very good opportunity for rebuilding on some entirely new plan.' He summarized the mixed results of the scheme and concluded: 'Net benefit has undoubtedly resulted. But it is a question whether an equal benefit might not have been gained in some gradual, less disturbing and less costly way.'[1] Indeed, this had become a general philosophy, and he was content that 'in spite of the poverty and drunkenness, domestic uncleanliness, ignorance and apathy that still prevail, things are surely making for the better . . . It is all a process of tinkering. Improvement is not coming structually from a Haussmann or socially and industrially by the light of master minds, nor is attempted by the dangerous road of revolution.'[2]

In his attachment to decentralized piecemeal reform Booth was at one with the views of the Royal Commission. His volumes with their close attention to local detail, and absence of organization around general themes, reflected not just muddle and lack of theory but also the manner in which he expected his work to be useful. Within this context, however, by 1900 there had been a change of emphasis, and Booth had emerged as a leading advocate of a suburban solution. This was to involve not an organized exodus of the poor, as he had once envisaged in his scheme for labour colonies, but a voluntary decentralization of the artisan.

In classifying the population of London, Booth began with Class A, 'the lowest class of occasional labourers, loafers and semi-criminals'. He concluded, contra Beames, that 'the hordes of barbarians of whom we have heard, who, issuing from their slums, will one day overwhelm modern civilisation, do not exist'.[3] Apart from the question of quantity, however, Booth's judgements differ little from those of his predecessors: 'This class tends . . . to herd together, and it is a tendency which must be combatted, for by herding together they –

both the quarters they occupy, and their denizens – tend to get worse. When this comes about destruction is the only cure.'[4]

Booth was, indeed, quite happy to use the concept of 'Alsatia': 'This semi-criminal class had its golden age in the days when whole districts of London were in their indisputed possession. They mainly desire to be left alone, to be allowed to make an Alsatia of their own.'[5] Even at the end of the century, a drop in the number of 'disreputable poor' in Westminster was ascribed to 'a consider-able exodus, partly to Lambeth, but mainly to Battersea and to Fulham where a new Alsatia is being found'.[6]

The isolation of the poor was reflected more generally in Booth's famous notion of 'poverty trap'. It was 'a very general rule that groups of poor streets, when cut off from communication with the surrounding district and so lacking the guarantee which through traffic provides, tend to become disreputable'.[7] This had arisen within the building process, and even parish boundaries might have some importance, since 'often it is on their line that streets are cut short so as not to "go through", and where this is the case there is a tendency to bad conditions of life, moral as well as physical. The wretched strip to the north and east of Tabard St. is a flagrant example of this tendency.'[8] He also stressed physical obstacles such as railway, river, and canal. Thus, 'in Battersea poverty is caught and held in successive railway loops . . . this is one of the best object lessons in poverty traps in London.'[9]

Although Booth reserved the notion of 'Alsatia' for the haunts of Class A, the 'poverty traps' were less closely identified in terms of class divisions, although they certainly also included the streets marked dark blue on his maps, which he associated with the very poor of Class B. These poverty streets may also be identified with the slum, an identification which is usually not explicit but follows from Booth's general tendency to match low conditions of life and habitation. Moreover, George Duckworth, one of Booth's assistants, in a much later article recollecting his work, was to define the slum as 'a street, court or alley which reflects the social condition of a poor, thriftless, irregularly employed and rough class of inhabitant'. He also surely reflects the thought of the master in his belief that 'the site of the slum is generally a pocket off a main street or a nest of streets where through traffic has been made impossible'.[10]

Duckworth's comments on such blocks of poverty or slum are also of interest. There was, for instance, 'hereditary poverty as in Westminster where the roots of poverty and crime have their origin deep seated in history in connec-tion with the Abbeys benefactions and its right of sanctuary for criminals'. There was also 'parasitic poverty . . . where a race of poor dependents has grown up encouraged by the indiscriminate giving of the well-to-do'. However, apart from certain vicious elements, slums, says Duckworth, 'are apt to be associated with those whose work, while it is normally unpleasant or irregular, is yet definite work and honest work. The slums tenanted by these last used to be described in our notebooks for Mr Booth concisely as gasworks poverty, coster poverty, dock and waterside poverty, so far as men were concerned, and jam works and laundry poverty particularly as affecting women.'[11]

It will be evident that here too there is little to distinguish Booth from his predecessors. His work in relation to slum and poverty blocks is but an extension of theirs, using superior methods. There is, however, a change of emphasis, for without abandoning moral contamination as a cause, there is a greater stress on factors of employment. This development is not originated by Booth, being already reflected in the views of the Royal Commission. It does, however, form one of the points of departure from which Booth was able to derive a much larger conception of the geography of poverty, a conception only faintly present in the mid-Victorian period. Looking at his poverty maps, Booth on the one hand sees a mosaic structure resulting from disparate factors:

> In Southwark much dark blue and black in courts and streets caused by a large demand for rough outside labour, by a traditional reputation for vice, by incomers from the Drury Lane neighbourhood, and by the presence of the lowest class of prostitutes, bullies and racecourse thieves; in Bermondsey much dark blue due to Irish waterside labourers and fish curers, and a rough class employed in wharves, markets and leather factories.[12]

However, these same patterns can be presented in a different way:

> With some exceptions, which bear explanation, south London poverty everywhere lightens as we recede from the river. There is a deplorably low level in all parts which lie near the sources of work. [Hence] this huge population . . . is found to be poorer ring by ring as the centre is approached.[13]

The need for the poor to live close to their work thus becomes the key not just to a local geography but to that of the whole city. The central location of employment, together with the relative access of various classes to transport facilities, determines the pattern of rings in which poverty is concentrated in the inner belt. Such an analysis was even more clear-cut in the case of overcrowding which Booth depicted from the results of the 1891 census.

It must be stressed that the mosaic and ring structures were not alternatives, but simply different aspects of the city. Attention could, however, switch from one aspect to another. This is evident in Figures 4.1–4.3. The maps of poverty streets and of clearance schemes may focus on the discrete and fragmentary nature of these distributions, but there is an obvious clustering around the central core of the city. Similarly, although the map of overcrowding, drawn at a greater level of generalization, exhibits the ring tendency more clearly, it also reflects something of the mosaic structure of the city.

The City of London, which lies at the heart of the larger geography, was not itself specially marked by poverty or slums in the late Victorian period. These had mostly been driven out by the expansion of business activities and, it was alleged, 'to lower the poor rates of the City parishes'.[14] It was the immediate surrounds of the City which saw the greatest concentration of slums, and at the beginning of our period this surround was still complete, separating the City from the wealthier parts of the West End. The belt of slums between these two continents was subject to special pressures, and some of the most famous slums and clearance schemes ran north from Drury Lane through Holborn to Saffron Hill and Clerkenwell. By the end of the period poverty and slums had been squeezed out from a good deal of this district, and the old ring transformed into

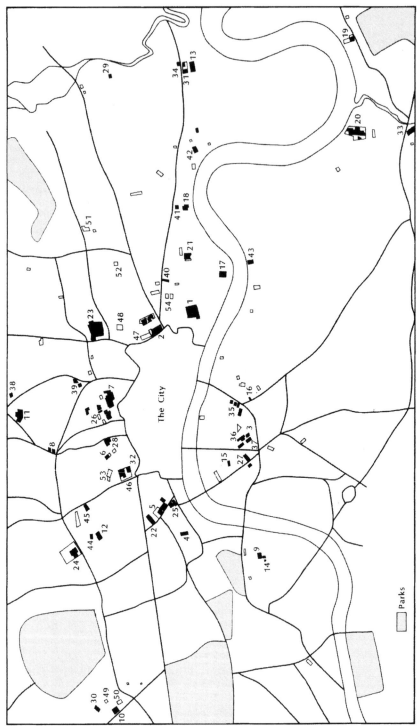

Figure 4.1 Slum clearance schemes and projects, 1875–1907. Black portions indicate completed schemes.

a horseshoe which extended around the commercial parts of the West End as well as the City. To the west the horseshoe was more broken, with isolated blocks in Lisson Grove (Marylebone) or Westminster. It thickened in the north from Holborn across Clerkenwell to Bethnal Green, and down the eastern side in Whitechapel. And although in the east poverty and slums again decreased away from the City, they still followed the river as far as Poplar. To the south, too, there was a clustering along the river from Southwark through Bermondsey and Rotherhythe to Deptford and Greenwich.

Difficulties in the implementation of Cross's Act contributed greatly to a switch of attention from slum block to slum zone. It became more apparent that the blocks were themselves embedded in a zone of competitive land use with relatively high land values. Equally, the relations between this zone and the suburban ring came back to prominence, notably in the suggestions of Alfred Marshall. However, although Marshall looked to the suburbs for a long-term solution of urban problems, he continued to emphasize the pressures on the inner ring, and in some ways simply transferred the analysis of the slum block upwards to a larger scale. The problems of the inner city were not so much the result of the modern economy and its urban structural effects, but an unfortunate result of historical persistence. Whereas factories were now avoiding urban congestion, workshops and domestic activities were 'not so ready to seek the fresh air. The causes of this are chiefly morbid, and their action is most conspicuous and calamitous in London.'[15] Here

> there are large numbers of people with poor physique and a feeble will, with no enterprise, no courage, no love and scarcely any self respect, whom misery drives to work for lower wages than the same work gets in the country. The employer pays his high rent out of his savings in wages, and they have to pay their high rents out of their diminished wages. This is the fundamental evil.[16]

Completed schemes
1 Rosemary Lane
2 Goulston Street and Flower and Dean Street
3 St George the Martyr
4 Bedfordbury
5 Great Wild Street
6 Pear Tree Court
7 Whitecross Street
8 High Street, Islington
9 Old Pye Street
10 Bowman's Buildings
11 Essex Road
12 Little Coram Street
13 Wells Street
14 Great Peter Street
15 Windmill Row
16 Tabard Street
17 Tench Street
18 Brook Street
19 Trafalgar Road
20 Hughes Fields
21 Cable Street
22 Shelton Street
23 Boundary Street
24 Churchway
25 Clare Market
26 Garden Row
27 Webber Row

28 Aylesbury Place and Union Buildings
29 Favonia Street Place
30 Nightingale Street
31 Providence Place
32 Brookes Market
33 Mill Lane
34 Ann Street
35 Falcon Court
36 Green Street
37 Gun Street
38 Norfolk Square
39 Moira Place
40 London Terrace
41 Queen Catherine Court
42 King John's Court
43 Fulford Street
44 Brantome Place
45 Prospect Terrace

Other projects
46 Holborn (1875)
47 Bell Lane
48 Great Pearl Street
49 Devonshire Place
50 Burne Street
51 Digby Street
52 Brady Street
53 Warner Street

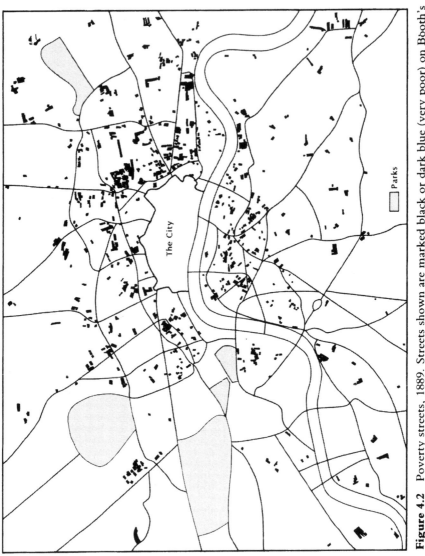

Figure 4.2 Poverty streets, 1889. Streets shown are marked black or dark blue (very poor) on Booth's Descriptive Map of London Poverty in Booth, *Life and Labour of London Poverty* in Booth, c. 1902–3 edn vol. V.

Parks

The City

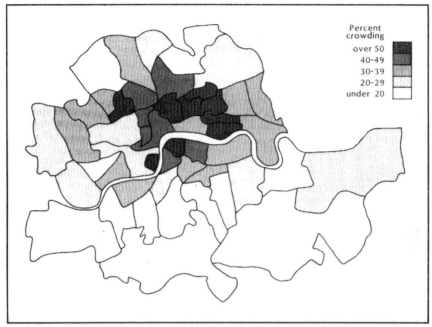

Figure 4.3 Overcrowding in London, 1891. The map shows the percentage of overcrowding according to the Booth standard of two or more persons per room. (Drawn from the Table of Districts arranged in order of Social Condition in Booth, c. 1902–3 edn, Final Volume.)

The solutions Marshall offered were hardly encouraging. Rigorous sanitary enforcement in 'proclaimed' areas would reduce cheap habitations, and ensure that 'a good many shiftless people who now come to London would stay where they are'. Also, 'wherever the dwelling houses of the poor were cleared away for any purpose, public or private, the requirements of conscience or of the law might in many cases be satisfied by handing over to a properly chosen committee money enough to move the displaced poor out into the country'.[17]

By the turn of the century, however, Booth was able to offer a much more optimistic scenario. His studies of London over the last decade had shown 'expansion in all directions following lines and laws so definite as to provide a stable basis for action'.[18] Expansion owed something to pressure from the centre, but was also dictated by choice. Choice in turn depended on transport, and this was the key to the whole process. The congestion of urban centres was a historical phenomenon which no longer reflected technical possibilities. 'Let anyone now design a city for four or five million inhabitants and how greatly it would differ in plan and structure from London. The possibilities of improvement . . . today rest entirely with improved means of communication.'[19] Thus, 'London's most urgent need is the removal of the trammels that now restrain movement and development.'[20] 'A new measure has to be applied to time and space in city life.'[21]

Booth emphasized, 'We want to see London spreading itself over the Home
Counties, not as an escape from the evil left behind, but as a development of
energy which will react for good over the whole area as it now exists, even in its
blackest and most squalid centres.'[22] 'If I advocate improved means of communication as a cure for these evils, it is because I believe that any direct attack
upon them is likely to fail.'[23] The new method of expansion would be indirect
because 'the movement of population might sometimes be from centre to
extremity, but would far more often pass from belt to belt; some in each belt
finding it convenient and natural to move one stage further on'.[24] Similarly,
expansion would have to filter down the social scale: 'However immoveable the
very poor may be, the same rule does not apply to men of settled employment
rather better paid.'[25] Change would be gradual, but relief immediate, expansion
acting like 'drainage on stagnant water-logged land'.[26] Indeed, going further,
Booth was prepared to offer a vision of the future in which the alternative
attractions of centre and suburb would be weighed according to family circumstances rather than class: 'The mere advantage of nearness to work will no
longer suffice, nor the mere escape from the pit.'[27]

Although Booth's scheme relied obviously on contemporary transport
developments such as the electrification of tramlines, it was not entirely a
technological solution. Technical potential needed also efficient administration.
In particular, transport monopolies needed to be broken down, and he became
chairman of a committee formed to press the view that 'a complete system of
transportation radiating from the congested centres, which shall be cheap, rapid
and owned by the London County Council, is a primary step towards dealing
with the housing problem in London'.[28] He was prepared to contemplate
extension of county boundaries, site value rating, and perhaps municipal land
ownership. However, all this was partly designed to forestall municipal housebuilding. For 'the futility of municipal action for the direct supply of dwellings
. . . is patent, and the dangers of this course, if pursued far, are very serious'.[29]

This is not the place to discuss the weaknesses of Booth's scheme. The object
is to trace the development of a new paradigm strategy, and this, I have stressed,
requires political plausibility, not rigorous logic or experimental proof. The
same general strategy was, of course, capable of extension in other political
directions, as will be shortly discussed. Moreover, when Booth came to present
his case to the Royal Commission on London Traffic, his impact was eclipsed by
the statistical presentations of the LCC, as organized particularly by E. J.
Harper. If we give first place to Booth, it is not because he was the most
important proponent of a suburban solution, or its initiator. It is because his
work was rooted more deeply than most in urban structure and systems, and
above all because it so amply covers a movement from one mode of thought to
another. Booth's own movement was not complete, and his 'final conclusions'
show him still deeply enmeshed in characteristically Victorian concerns:
'Improvement must be sought first of all in the deepening of the sentiment of
individual responsibility.' Cynicism and 'the ready acceptance of low standard'
were still prevalent difficulties. He added, ominously, 'This moral laxity applies
to all classes.'[30]

The chronology of clearance

Summarizing the reform efforts of municipal authorities in the Victorian period, Fraser writes, 'Involvement was liable to fluctuate between peaks of activity and troughs of inertia.' This was partly because the results of reform were often disappointing: the evils were always so much more deeply rooted, and the anticipated gains so much more elusive than had been expected. An important factor, though, was that reform brought increased costs, and a counter-movement of economy often involving change of municipal control. Yet 'the same public who cried out for economy were ambivalent and could soon adopt a quite different posture. As always support for municipal expenditure varied directly with the death rate.' Economies having been made, sooner or later the pressure of intolerable facts would push forward the reform movement, and so begin the cycle anew.[31]

The process was not entirely cyclical, being influenced by longer-term developments. Thus, in the late Victorian period, factors such as the effect on confidence of the Great Depression, and growth of 'socialist' ideas, helped to weaken the ideological blockage to reform created by a commitment to *laissez-faire*. More particularly, Fraser's work has been directed to the connections between public health reform and the development of municipal government itself. Changes in the electoral base, bringing a widening of democracy, had long-term if not straightforward effects. Growth of cities and their economies brought an expanding tax base. Above all, reform agitation stimulated local government reorganization, so that over the course of the Victorian period it was possible to speak of a revolutionary transformation in municipal government. This not only increased the potential effectiveness of municipal involvement but created institutional bases, such as the post of medical officer of health, from which reform measures could be further advanced.

The cyclical pattern of development can be clearly related to the progress of slum clearance in London. The sporadic nature of the programme is very evident, and indeed, of the £2 126 234 estimated net expenditure approved for clearance by the Board and the Council between 1875 and 1907, 68 per cent was allocated in only four years: 1876, 1890, 1895, and 1899. Following these surges attention switched to the more disagreeable aspects of clearance, compensating landlords and displacing tenants, before any rebuilding was achieved. Above all, actual expenditure mounted, reaching peaks, as recorded by capital transfers, in 1878–81, 1892–3, and 1898–1901 (Table 4.1). These were the years of agonizing reappraisal in which new approvals were restricted or even halted altogether. This expenditure then declined, and the stage was set for a potential renewal of activity.

The crucial importance of costs is also borne out in the attitude of the vestries. Whatever the motives for removing slums, and perhaps their inhabitants, from their areas, the vestries were certainly not prepared to spend their own money on such projects. There were, however, always plenty of candidates for Part I schemes. A further link with the Fraser model is that the replacement of the Board by the Council was undoubtedly prompted in part by public health considerations and was generally followed by more active and efficient government. The reputation of the London Council for reform was due in part to the

Table 4.1   LCC and MBW clearance expenditure, 1875–1905.

Year ending	Approved estimated net expenditure (£)	Net capital transfer (£)*
31 December 1875	54 400	—
1876	412 287	—
1877	117 800	30 000
1878	84 747	380 500
1879	—	300 000
1880	—	173 884
1881	—	500 003
1882	86 480	3
1883	—	38 532
1884	77 100	32
1885	—	—
1886	101 850	45 000
1887	—	50 176
1888	—	—
Subtotal	934 664	1 518 130
31 March 1890	—	45 000
1891	300 000	5 049
1892	3 700	24 465
1893	23 667	184 998
1894	32 300	91 097
1895	—	} 14 667
1896	268 150	
1897	14 107	12 287
1898	21 306	70 368
1899	10 333	139 275
1900	506 900	173 557
1901	—	124 744
1902	11 107	64 837
1903	—	1 505
1904	—	47 686
1905	—	−28 446
Subtotal	1 191 570	971 089
Total	2 126 234	2 489 219

Sources: MBW minutes and LCC accounts, *passim*.
*Transfer to Dwelling House Improvement Fund.

fact that compared to provincial cities, which had benefited from earlier reorganization, 'there was a lot of immediate ground to be made up in the metropolis'.[32] Again, the reassertion of the need for the Council in 1898 was followed by a further burst of activity.

However, the pattern of expenditure on slum clearance shows no tendency to rise in response to longer-term factors, and the efforts of the Council were certainly no greater than those of the Board. Between 1875 and 1885 the Board's capital expenditure on clearance schemes was £1 518 130, compared with £932 440 spent by the Council between 1889 and 1907. Such well-known figures receive all the more emphasis when set against current rateable values for London which increased by 86 per cent between 1876 and 1906. As a percentage

Table 4.2 Clearance expenditure and rateable values.

	1 Capital expenditure (£)	2 Annual rateable value (£)	3 Ratio 1:2
1876–80	884 384	23 240 000	3.80
1881–5	538 570	27 629 000	1.95
1886–91	205 225	30 716 000	0.47
1891–6	309 953	33 004 000	0.94
1896–1901	467 125	35 793 000	1.30
1901–6	51 565	39 643 000	0.13

Note: 1876–80 and 1881–5, January–December; 1886–91, January–March; other years to 31 March.

of current rateable value, the expenditure of the Board in 1875–80 was three times that in any quinquennial period under the Council (Table 4.2).

It is true, of course, that most of the Board's expenditure took place before it was aware of the actual costs of clearance, whereas the Council's schemes were frequently less costly than originally envisaged. Certainly, in terms of approved estimated expenditure there is less contrast between the Board and the Council, and the Council did much more in the 1890s than the Board had done in the 1880s. Even so, the efforts of the Board in the early years of Cross's Act were never again equalled in real terms, and undoubtedly there is a break in experience here, for the politics of clearance were never the same after the Board's debacle. As suggested in Chapter 3, the work of the Council in the 1890s took place against a background of retrenchment.

Table 4.3 LCC capital expenditure, 1889–96.*

Item	Value (£)
street improvements	577 770
main drainage	884 084
parks and open spaces	584 538
bridges, tunnels, and ferries	1 186 405
clearance schemes	360 002
artisans'dwellings	180 401
lunatic asylums	720 786

Source: LCC 1897. *London statistics,* vol. VI, pp. 630–1.
*31 March in each year.

To a large extent, indeed, this must be explained by factors particular to slum clearance, for the Council was not generally adverse to capital expenditure. Table 4.3 shows the small proportion of total capital expenditure taken by clearance schemes compared with other 'non-remunerative' ventures such as street improvements, main drainage, bridges and tunnels, asylums, and even parks and open spaces. The expenditure on artisans' dwellings was, of course, expected to be remunerative. In the face of such figures it cannot be argued that lack of progress in clearance was simply a problem of cost.

As local government expenditure in this period fell exclusively on the rates, it was always liable to be opposed on distributional grounds. This opposition might involve proposing some other source, such as ground rents or imperial taxation, on which costs might fall. The politics of the taxation of land values, and of the land question in general, are very relevant here. Although a special connection with ground values could be established for some other expenditures, notably street improvements, it was particularly strong in the case of slum clearance. This strength may partly explain why clearance was also to be an early target for the alternative Conservative prescription of subsidies from imperial taxation. Slum clearance was even more obviously susceptible to the charge that its costs should fall not on the ratepayers but on the special group of persons affected. This meant more than a view that each should be responsible for his own, for in the case of the slum the questions of 'responsibility' and 'negligence' were posed in a highly charged way. It could more easily be argued that the costs of remedial action should fall on slum tenants and landlords. Slum clearance, as the action of the responsible community against the irresponsible, should be paid for by the latter. To do otherwise would be to encourage malpractice.

The increased politicization of local government represented by the replacement of the Board by the Council heightened these distributional questions. Although blockages caused by opposition to the principle of intervention were diminished, new blockages were introduced by political attitudes to land and landlords. Ultimately, these attitudes not only produced increased resistance to clearance but also played an important part in the adoption of a fundamentally different policy towards the slum.

To the complex politics of the land question must be added increased political pressure from the working class, particularly from organized labour. Byrne and Damer argue for a very close relationship between housing policies and class politics. Regarding the potential use of the 1890 Act. They say there was 'a desire to confine such use to a reduction of over-supply (if that) on the part of the urban bourgeoisie, and a desire to engage in direct municipal provision on the part of the emergent working class organisations'.[33] This analysis is in turn derived from Foster's review of developments in London. For him, '1898 was the one year in the late nineteenth century which saw a policy initiative which cannot be directly explained in terms of the housing cycle and sectional interests. It marked the beginning of the LCC's one major housebuilding programme before 1914.' By contrast, 'very crudely, but in ugly conformity with the propertied interests it represented, Parliament became concerned with housing when there was a glut of empty houses not a shortage'. In fact, 'parliamentary activity coincided with crises of profitability'.[34]

It has been established that the index of empties follows the long building cycle rather than the shorter trade cycle with its peaks and troughs of unemployment, although the latter still has some effect. Parry Lewis argued: 'When empty building reached a certain minimum percentage, the normal movement and growth of population ensured that building would be profitable. But as building proceeded, and eventually overtook demand, so that empty property became more abundant and rents began to fall, so it would become less profitable.'[35] Within our period the points of crisis of profitability thus indicated are the late 1860s and the mid-1880s, each following building booms of varying intensity.

Foster links this chronology particularly to the advent of the Royal Commission of 1884–5, and also to the legislation of the late 1860s. Yet he omits from consideration altogether Cross's Act of 1875, which would seem most relevant to this argument, but was passed at a time when empties were at a trough. Moreover, of the events that seem to fit the Foster pattern, Torrens's Act of 1868 was notable for its lack of immediate application, and no satisfactory connection is made out between the Royal Commission and actual legislative and practical effects. Also important, surely, is the incidence of adoption of slum clearance programmes, which is rather more definitely related to low points in empties than high. This is true of both the peak years in 1876 and 1899, but the mid-1880s saw notably little clearance. It would not be difficult, indeed, to supply a logical link for such a connection, since crises of profitability would create increased resistance to the costs of clearance.

This does not mean that 'reducing oversupply' was not a motive for clearance among some groups. The incidence of legislation, or of the adoption of clearance programmes is, however, clearly affected by other factors which at best mask its effects and make it impossible to discern any precise chronological relationships. A difficulty then presents itself in that the Foster thesis particularly relies on this form of argument, taking the view that the actual reasons given for promoting clearance schemes and other measures were mere camouflage concealing the real intentions which can only be discerned in effects.

Similar difficulties present themselves in respect of an apparent relationship between the adoption of clearance schemes and periods of boom and low unemployment. If, taking the contrary view to Foster, one wished to present clearance as a genuine reform movement, a possible mechanism for this connection is suggested by Thompson's assessment: 'Trade union development closely reflected the changing economic situation. Between 1887 and 1891 there was a strong recovery in the economy, and a notable expansion of trade unionism. This was followed by a sharp depression 1892–5 which seriously weakened the unions; a recovery 1896–9 accompanied by a revival of trade union militancy. . . .'[36] Could it be that working-class pressure, swelling up in periods of relatively high employment, exerted a positive influence on clearance adoption, either directly or through the election of progressive elements? Here again, however, an obvious caveat presents itself, for it was in such periods of boom that expenditure for clearance was, for whatever motive, more likely to be released.

It seems to me that the Foster thesis errs in presenting slum clearance as a sharply focused policy, narrowly motivated, with clearly divided forces marshalled in its support and attack. On the contrary, it was rather a loosely-knit bundle of policies, which attracted support and attack for a variety of motives. Usually, indeed, a mixed view of the strategy was taken, and probably relatively few unequivocally supported or rejected the whole range of ingredients and consequences which such a policy implied. Moreover, a unitary view of policy, by presenting no rough points or areas of potential development open to exploitation by various interests, both reduces the scope for unintended or undesired effects and reduces the possibilities of effecting change within the overall framework of the existing strategy.

Such criticisms apply to some extent even to much more acceptable short descriptions of clearance, such as Merrett's view that it was 'done in order to physically destroy nests of disease and crime. The building of new dwellings on the vacated sites (or elsewhere) basically served to legitimise the destruction which preceded it.'[37] It has to be recognized that a rehousing obligation could not be included without having a major effect on the rest of the programme, and this was particularly so in relation to the designation of sites and compensation. Indeed, considering its impact on costs, it is apparent that rehousing, though legitimizing clearance, also served to prevent it. Moreover, rehousing has, in varying degrees, to be taken as an objective in its own right. The provision of sites in central London for the dwellings companies was a prong of the original Act. Later, clearance opened the way for municipal building, and by the end of the century the building element was an important consideration in determining whether clearance should take place.

The connections which undoubtedly exist between the 'internal' cycle of adoption and rejection of schemes and the 'external' economic cycles seem best explained by the fact that boom helped to release the purse strings, whereas slump brought increased resistance to expenditure. There are no necessary implications as to particular motives. The existing strategy benefited from the feeling released in more comfortable times that 'something must be done', if only by way of example. It did so because it offered something for a wide variety of interests, points of attachment which they could seek to develop. Thus the more radical element on the first Progressive Council sought to make progress not so much through Part III as through the scope that seemed to be offered for an attack on compensation and property values by greater emphasis on the principles of Torrens's legislation.

It was only from about 1892 that an alternative strategy built around Part III began clearly to emerge and to form a rallying point for opinion. Even so, its progress was at first slow, not just because of the opposition of sectional interests but because, like its predecessor, it was made up of a linked structure of various elements which varied individually in their acceptance to different groups. The new paradigm involved, for example, an elimination of subsidy, a very optimistic view of filtering, and a switch of attention away from the areas of greatest housing stress and from those in worst housing difficulty. The kind of approach adopted by Foster, Byrne, and Damer not only leads them to give too black a view of the old strategy but also to give too favourable a view of the new.

The suburban solution and its alternatives

The logic of the Progressives suburban strategy was best developed in the writings of William Thompson, initiator of the Richmond scheme and a leading propagandist for this approach. Its centrepiece was an emphasis on overcrowding, but this concern, which ran back to the Royal Commission and before, had necessarily to be developed in new directions. Overcrowding was now discussed in terms of the notion of 'house famine', and from this it followed that 'a very large increase in the supply of all kinds of sanitary housing . . . is the only measure likely to have a material effect'.[38] If, however, attention was concentrated on extreme cases of overcrowding, and thus on the very poor, the traditional argument would seem to follow that such persons needed to be housed close to the city centre. Thompson countered this by emphasizing 'the widespread nature of the evil and the fact that it affects not the indigent poor alone, but the great mass of working people of all grades'.[39]

Existing clearance and rehousing policies had entailed large expenditure on 'compensating slum owners and forcing up the market value of bad property by diminishing the supply . . . the money cost of continuing this policy will be ruinous both to the ratepayers and to the working classes who reside in such areas'.[40] Moreover, 'while the plague spots have disappeared and the disease has lessened in local intensity, the general sanitary condition of the district as a whole has not materially improved for any length of time'.[41] Forcing up sanitary standards in existing dwellings during a house famine could only increase rents and rebound to the advantage of the landlord so that 'the tenants in many cases are the strongest opponents to sanitary inspection and improvement of their houses'.[42] A large reduction in overcrowding was, therefore, a *sine qua non* of all other improvement.

Overcrowding, lynchpin of a whole nexus of evils, was itself the product of a failure of private enterprise to supply sufficient housing, a failure not confined to the inner city. Increased supply through Council building was the remedy; it would react on existing conditions through competition and filtering. Attention was switched from the fact that the poor could not move to the fact that others could. Given sufficient commitment, a radical change in the whole structure of the city could be envisaged, for if 'half the workers could be induced to leave the congested districts of London, exorbitant rents would fall, overcrowding would be diminished, and the health of the people enormously improved with little or no cost to the rates'.[43] There are obvious parallels with Howard's more famous vision of 1898:

House property in London will fall in rental value and will fall enormously. Slum property will sink to zero, and the whole working population will move into houses of a class quite above those which they can now afford to occupy . . . these wretched slums will be pulled down and their sites occupied by parks, recreation grounds and allotment gardens. And this will be effected not at the expense of the ratepayers but almost entirely at the expense of the landlord class.[44]

Thomspon recognized that for decentralization to work it was 'essential that cheap and rapid transport by electric tram should be established'. However, 'cheap transport alone will simply mean an increase in suburban rents and land values.' 'If private enterprise puts up a colony of cheap houses, it will simply mean a new suburban slum.' 'The organised dispersion of population by municipal action is the only practical and satisfactory remedy for present evils.'[45]

This meant more municipal housing and municipal transport, but above all more control of land.

Thompson's strategy was, indeed, ultimately a land strategy, the essence of which was the use of municipal powers to obtain land close to agricultural prices, so ensuring that the 'unearned increment' arising from urbanization was not privately appropriated. Municipal control of transport was a vital element here, because it could be directed to the opening up of areas which had been transport poor and therefore retained low land values. Land purchase in advance of speculative price rises was essential. In 1900 Thompson argued that initially 'the purchase of say 10 000 acres of such land within a 12 mile radius of the City of London might reasonably be effected . . . the colonies thus established need not be confined to one district and ought to be bought at a reasonable, if not a cheap price'.[46] In this way cottages could be built at sufficiently attractive rents, and the great revolution accomplished, as Howard had dreamed, 'not at the expense of the ratepayer but almost entirely at the expense of the landlord class'.

Municipal building could thus continue to 'set up a standard of a decent sanitary home that a working man might reasonably expect'.[47] Set in the suburbs this would be of cottage style, and the contrast with the tenement blocks of the inner city was an obvious point. For Thompson, the latter were 'more or less objectionable' and even the best of them were injurious to health.[48] Likewise, the possibilities of lower-density development and improved environmental design might also become part of the argument. The core of the strategy lay, however, in the pledge to increase housing supply, and the suburban orientation of these policies depended ultimately on financial logic rather than environmental preference, though if the two went together so much the better.

In 1903 the *Municipal Journal* described the Progressives and Moderates as 'the parties of action and inaction in regard to municipal affairs'.[49] This was not inappropriate, and certainly the Progressives had prospered electorally when the powers of the Council were at stake. Through much of the 1890s, however, a breach within the party had seemed to oppose action on the one hand and antilandlordism on the other, particularly in the field of housing. One of the great benefits of the Thompson approach was that these two currents could be reconciled, and the powerful land question turned fully to the party's advantage. Moreover, in addition to land acquisition, the strategy required an extension of municipal trading in the fields of both housing and transport. These were all closely linked policies which would serve together to break the monopoly of the landlords.

Clean, efficient municipal estates served by clean, efficient municipal tramways represented a politically useful vision of the future. It appealed directly to the most powerful section of the working-class electorate on whose support the Progressives increasingly depended, to those 'respectable labourers and artisans who have as much necessity for a decent home as those in a lower grade'.[50] Politically, however, the indirect effects of Council building were equally important. Rallies called for 'the speedy erection by the Council of good homes for the people in sufficient numbers to bring down rent'.[51] In this way, Council building could be connected with the rent agitation of the period, expressed also in demands for fair rent courts. More generally, the council could claim to be contributing towards a solution of inner-city problems, whereas the responsibility for those problems was attached more firmly to the landlords.

The apparent simplicity of this model was clouded when the location of Council building became apparent. The geographical isolation of such projects was an obvious weak point in electoral presentation. Moreover, there was a logical contradiction in that the failure of filtering was so strongly emphasized in opposition to Part I schemes. Now, the same degree of filtering from one class to another was compounded by a necessity for geographical filtering. For this reason, many Progressives clung to the hope that Part III policies could be applied without the consequence of suburban location. This was unrealistic, however, because such location was a logical outcome of a policy which aimed to work with the market.

The matter of subsidies was one which posed clear dangers for the Progressives. In the period of disillusionment and reappraisal of Part I policies the question had arisen whether 'it is so very unreasonable to rehouse out of the rates some of the people evicted'.[52] Knowles reported, however, in 1899 that 'the proposal causes even ardent Progressives to hold up their hands in horror. The Progressive leader has declared that such a policy would wreck his party.'[53] This was probably not an exaggeration. In the current political context a subsidy of this kind could only be justified where rehousing obligations were incurred, as in Part I schemes, and would thus simply compound the already fierce opposition to those projects. The whole basis of an approach to housing through land reform or municipal trading was that problems could be solved without subsidy and by acting within the constraints of the market. Any overt subsidy for building was ruled out not only by political opposition but by the inherent objectives of powerful sections of the party.

As the logic of the Part III strategy unfolded, there appeared the beginnings of a considerable reversal in previous party positions. It had always been the Moderates who had placed most emphasis on a suburban solution, whereas the Progressives had emphasized that they alone fulfilled their responsibilities in clearing the slums and housing the working classes near their work. Changed alliances were revealed in the discussion over rehousing at Clare Market when 'Mr Sidney Webb deprecated the proposal to use such valuable commercial land for the purpose of erecting dwellings for the working classes', and Colonel Rotten, Moderate, 'argued that it was vitally necessary that most of the poor should be housed near to their work . . . adding that it was . . . a wrong policy to spend huge sums of money in housing people at such places as Tottenham or Norbury'.[54]

There can be no doubt that the Moderates were considerably disconcerted by the way in which their opponents had turned the suburban solution to their own advantage. The incorporation of municipal house building for general needs was a particular anathema. In the short run, however, opposition could take the form of simple reaction, and the eventual election victory of 1907 was based on a classic appeal to economy. The task of building up some plausible alternative to the Part III strategy thus fell to administrations outside London, principally in Birmingham and Liverpool. It was to these models that the Moderates were to look when opinion again moved round the cycle from inaction to action and demanded that something must be done.

One possibility was to embrace the suburban solution, and to find some means by which the Council could be seen to actively promote it without actually engaging in housebuilding. This would have the advantage of switching attention away from the slums, a focus which still offered disagreeable possibilities in the form of attacks on compensation, subsidy proposals, and even a prospect of rehousing obligations being imposed on private developers. A. J. Balfour certainly took the view that 'the remedy for the great disease of overcrowding is not to be found in dealings, however drastic, with insanitary areas', and showed some attraction to Booth's solution based on greater municipal involvement in transport.[55] In Birmingham, John Nettlefold was one of the first to grasp the political possibilities of using town planning as a means of reconciling municipal initiative with private development. His approach, however, though avoiding municipal building, still required a belief in big city government, incorporation of suburbs, and even municipal land purchase, and was difficult for the Moderates to apply in the context of London government.[56]

The other possibility was to re-focus municipal attention on the slum and leave the suburbs to private development. Here Nettlefold again offered a model, based on a development of the Part II policies which Moderates had once favoured as an alternative to Part I schemes. This involved compulsory rehabilitation combined with selective demolition in improvement areas. However, such action in overcrowded districts inevitably involved considerable displacement of population and was thus logically linked, however effectively, to the other prong of Nettlefold's policy. Without this, the actions of the Council would appear in a more negative light. There was the further difficulty that rehabilitation of this kind could only be done with the co-operation of the London boroughs within whose province these powers lay. Such action was therefore not to be adopted, although there was a brief flirtation in 1912 when Nettlefold was consulted over the development of the Brady Street area of Bethnal Green.[57]

The one unequivocal power which the Council did possess was to effect clearance and rehousing schemes, whether under Part I or Part II. Here a relevant line of development was opened up by the policies of Liverpool Corporation. These had been foreshadowed in the 1880s, but were first implemented from 1896 when it was decided to 'go forward with housing schemes, restricting the use of the dwellings to the persons actually turned out'. This, it was claimed, 'satisfied the builder . . . because it ensured . . . that the new dwellings . . . should be for the poorer classes, for whom they could not and did not want to provide'. On the other hand, 'it satisfied those who desired that the occupants . . . be rehoused in or about their old habitations, near their shops, their places of worship, their schools, and above all in the neighbourhood in which they earned their livelihood'.[58] In 1902 the Council, having enumerated nearly 10 000 unfit houses in the city, planned to demolish these at a rate of 700 a year.

Our concern here is not with the origins of this policy, or its actual effects, but with the manner in which it might be built up as an alternative to the Part III strategy.[59] At the national scale this appeared clearly in the Commons debate on housing in 1903. Taylor, the Liverpool housing leader, maintained that central government should 'not only require that adequate accommodation should be provided to replace the houses that were pulled down, but should go further and demand that the accommodation should suit the capacities, the pockets and the conditions of those who were displaced from slums which the municipalities destroyed'. This would 'bring a little nearer . . . the practical solution of the slum problem which was the real problem for the municipalities to deal with'. He was followed by Keir Hardie who argued for Labour that 'instead of playing the part of scavenger to the private housebuilder, and taking charge of the refuse, out of which they could not make a profit, the Local Government Board should insist on the local authorities building homes which would cater for the better class of artisans. By that means housing vacancies would be created right down to the lowest slums, and the competition of the municipality would tend to reduce rents all round.'[60]

These developments would simply seem to resurrect one variant of the old paradigm to counter the new. Certainly, Liverpool's policy was initially based on building down and relied on the claim that 'Liverpool was the first municipality in the world which, without any loss to the ratepayers, had provided dwellings for its poorer inhabitants'.[61] Later, however, standards were raised somewhat, and Thompson in 1907 maintained: 'Reckoning the land at its full value the rents are subsidised by 2s. 6d. per room per week, while if the building cost alone is reckoned the subsidy is 7d. per room per week.'[62] By 1913 Taylor was happy to claim that 'the mere cost of building would prevent these people being housed at an economic rent'.[63] For this claim had meanwhile formed the basis for the strategy worked out by the Unionist Committee set up to counter the Liberals' 1909 Housing and Town Planning Act and Lloyd George's assault on land. Incorporated in the Housing Bill presented to the Commons by Boscawen in 1911 it envisaged national government subsidies for rural cottage building and urban Part I schemes.

On the re-election of the Moderates in 1910 Boscawen had been made housing chairman. His work with Taylor on the Unionist Committee provided a link with the Liverpool strategy and led to the first London-wide survey of unfit housing and the adoption of the large Tabard Street scheme. The higher costs of redevelopment in London meant, however, that there was no commitment to rehouse the displaced and the project was justified in very traditional terms: 'The effect of removing these rookeries from the heart of London is ample justification for the expenditure in view of the immeasurable benefit which will result to the health and well-being of the community.'[64]

Also following tradition, the Moderates had once more drawn back from further clearance by 1912. More generally, however, the Liverpool programme did offer a few points on which a new paradigm based on clearance, but directed rather more to the well-being of slum populations themselves, might in future be built. These included a definite programme based on comprehensive survey, subsidy, and above all the claims that there had been a 'social transformation' in which mortality had been reduced and the people were 'healthier, happier and stronger'.[65]

Notes

1   Booth, C. 1889–1903. *The life and labour of the people in London*, 1902–3 edn. Religious Influences Series, vol. II, p. 68. London: Macmillan.

2 ibid., p. 65.
3 Booth, C., Poverty Series, vol. I, p. 39.
4 ibid., p. 70.
5 ibid., p. 174.
6 Booth, C., Religious Influences Series, vol. III, p. 81.
7 ibid., p. 120.
8 Booth, C., Poverty Series, vol. I, p. 265.
9 ibid., p. 192.
10 Duckworth, G. 1926. The making, prevention and unmaking of a slum. *Journal of the Royal Institute of British Architects*, vol. 33, p. 328.
11 ibid., p. 328.
12 Booth, C., Religious Influences Series vol. IV, p. 163.
13 Booth, C., Poverty Series, vol. I, pp. 263, 276.
14 Denton, W., 1861. *Observations on the displacement of the poor by metropolitan railways and by other public improvements*, p. 23. London: Bell & Daldry.
15 Marshall, A. 1884. The housing of the London poor. *Contemporary Review*, vol. XLV, p. 225.
16 ibid., p. 226.
17 ibid., p. 229–31.
18 Booth, C., op. cit., Final volume, p. 15.
19 Booth, C. 1901. *Improved means of locomotion as a first step towards the cure of the housing difficulties of London*, p. 23. London: Macmillan.
20 Booth, Final Volume, p. 189.
21 Booth, C., *Improved means*, p. 23.
22 Booth, C., Final Volume, p. 205.
23 Booth, C., *Improved means*, p. 10.
24 ibid., p. 17.
25 ibid., p. 15.
26 ibid., p. 17.
27 Booth, C., Final Volume, p. 208.
28 Booth, C., *Improved means*, p. 7.
29 Booth, C., Final Volume, p. 190.
30 ibid., p. 210.
31 Fraser, D. 1979. *Power and authority in the Victorian city*, pp. 69–70. Oxford: Blackwell.
32 Kellett, J. R. 1978. Municipal socialism, enterprise and trading in the Victorian city. In *Urban history yearbook*, p. 39. Leicester: Leicester University Press.
33 Byrne, D. and Damer, S. 1980. The state, the balance of class forces, and early working class housing legislation. In *Housing construction and the state*, p. 61. London: Political Economy of Housing Workshop.
34 Foster, J. 1979. How imperial London preserved its slums. *International Journal of Urban and Regional Research*, vol. 3, pp. 93–114.
35 Lewis, J. Parry 1965. *Building cycles and Britain's growth*, p. 130. London: Macmillan.
36 Thompson, P. 1967. *Socialists, liberals and labour: the struggle for London 1885–1914*, p. 43. London: Routledge & Kegan Paul.
37 Merrett, S. 1979. *State housing in Britain*, p. 30. London: Routledge & Kegan Paul.
38 Thompson, W. 1903. *The housing handbook*, p. 7. London: National Housing Reform Council.
39 ibid., p. 1.
40 ibid., p. 4.
41 ibid., p. 8.
42 Thompson, W. 1900. Powers of local authorities in the house famine and how to relieve it. In *Fabian Tracts*, no. 101, p. 19. London: Fabian Society.
43 ibid., p. 25.
44 Howard, E. 1898. *Tomorrow: a peaceful path to real reform*, p. 44. London: Swann Sonnerschein.
45 Thompson, W., 1900, op. cit., p. 25.
46 ibid., p. 23.
47 Thompson, W., 1903, op. cit., p. 12.
48 ibid., p. 67.
49 *Municipal Journal*, 30 Oct. 1903, p. 959.
50 Thompson, W., 1903, op. cit., p. 50.
51 *The Builder*, 1 Sept. 1900, p. 199.

52 *Municipal Journal*, 20 Apr. 1900, p. 305.
53 Knowles, C. M. 1899. *The housing problem in London*. Housing Reform Union Pamphlet no. 81, p. 13.
54 *Municipal Journal*, 16 May 1902, p. 410.
55 Balfour, A. J., House of Commons speech May 1900, quoted in Booth, *Improved means*, p. 1.
56 Nettlefold, J. 1908. *Practical housing*. Letchworth: Garden City Press; see also Sutcliffe, A. 1974. A century of flats in Birmingham 1875–1973. In *Multi-storey living: the British working class experience*, A. Sutcliffe (ed.), pp. 186–9. London: Croom Helm.
57 LCC Minutes HC 11 Dec. 1912 (17).
58 City of Liverpool Housing Committee 1913 *Artisans' and labourers' dwellings and insanitary property*, p. 7.
59 On this see Taylor, I. 1974. The insanitary house question and tenement dwellings in nineteenth century Liverpool. In Sutcliffe A. (ed), op. cit., pp. 41–83; and Pooley C. 1985. Housing for the poorest poor: slum clearance and rehousing in Liverpool 1890–1918. *Journal of Historical Geography*, vol. 11, pp. 70–89.
60 *Hansard*, vol. 120, HC Deb., 4s., 2 April 1903, cols 941–5.
61 Ibert, L. 1899. Pioneers in housing. *Economic Review*, vol. IX, p. 458.
62 Thompson, W. 1907. *The housing handbook up to date*, p. 115. London: National Housing Reform Council.
63 *Hansard*, vol. 51, HC Deb., 5s., 11 April 1913, col. 2269.
64 LCC Minutes 8 Nov. 1910 (15).
65 Aldridge, H. R. 1923. *The national housing manual*, pp. 275–6. London: National Housing and Town Planning Council.

Part II

PROPERTY AND COMPENSATION

5 Compensation and land use

This chapter is designed to deal with the principles and practice of slum property compensation with special reference to the nature of physical structures and land use. These matters cannot, of course, be treated apart from the system of land and property ownership, but the latter question is taken up more fully in Chapter 6. It then leads on to a more general discussion of relationships between slum property compensation and market values.

Torrens, Cross, and compensation

The twin chains of reasoning which led to the Torrens and Cross Acts intertwine through all the related matters of property condemnation and compensation. They were not present simply in alternative pieces of legislation, but constantly recur in the struggles which took place over the extent of clearance or amounts of payment. This was notably the case when under Cross's legislation, originally framed to shift responsibility from the landlord to the state, more stringent compensation provisions were gradually introduced designed to shift it back again. Yet the Torrens principle, the responsibility of the individual landlord, was capable of being developed as a defence of property, just as the underlying theme of Cross's legislation, that conditions were beyond the capacity of individual landlords to remedy, had many radical implications.

Under Torrens's Act property first reported by the medical officer was subject to further assessment by surveyor, vestry, and magistrate. The emphasis on individual responsibility ensured attention to each particular case. At the same time, however, these procedures had acted to prevent clearance. It was a common complaint of medical officers that 'the worst part of the Act is that it refers the matter to the surveyor . . . the surveyor reports and the owner . . . patches them up in some way and the evil remains'.[1] Apart from the surveyor, vestry members and magistrates were also blamed for lack of progress. Even under Part II of the 1890 Act clearance orders were notoriously difficult to obtain. Between 1892 and 1898 a total of 788 were given in London, which resulted in 229 demolitions, 219 improvements, and 340 cases of no further action.[2]

It was often suggested that personal connections with slum property were responsible for this dilatory progress, as in the case of vestry members at Clerkenwell publicized by the Royal Commission.[3] Some general sympathy for the owners on the part of the officials concerned in the enforcement of the regulations was certainly a necessary condition. Beyond this, however, were factors which could be called up in defence of house landlords. It could be, for example, that individual sites were 'too small to contain new buildings of any commercial value', or that the unfit condition of houses was 'due to original structural defects and not to neglect to keep in repair or to provide proper sanitary arrangements'.[4] Remedial action might therefore be confined to those features thought reasonably within the landlords capacity to alter.

The key rôle given to the medical officer in Cross's legislation was partly designed to attach the measure firmly to public health rather than general town improvement. In other respects, however, it was a step taken to accelerate change. The surveyor and magistrate were bypassed completely, and each medical officer was given direct access to the executive authority. A medical officer was to make a representation when he considered that

> any houses courts or alleys within a certain area . . . are unfit for habitation, or that diseases indicating a generally low condition of health have been from time to time prevalent . . . that such prevalence may reasonably be attributed to the closeness, narrowness and bad arrangement, or the bad condition of the streets and houses . . . or to the want of light, air, ventilation or proper conveniences, or to any other sanitary defects . . . and that the evils cannot be satisfactorily remedied otherwise than by an improvement scheme for the re-arrangement or reconstruction of the streets and houses.[5]

Representations were brief, formal affairs, and, given the support of the executive authority, they came under serious scrutiny only at the subsequent local inquiries. These began from the basis that the necessity for a scheme was established by experts and could only be challenged by contrary expert opinion. In these circumstances it is not surprising that a scheme supported by the executive authority was never wholly rejected. The necessity for a scheme only became a matter of serious contention when local and executive authorities were opposed, and even then it was not until the advent of Part III policies that this was ever successfully refuted. By far the most important private opposition came from the Duke of Bedford at Coram Street, but even that did not deflect clearance altogether.[6] Lesser landlords did not band together in any organized body and were mainly concerned to obtain more favourable treatment for their own property, by exclusion from the scheme, or movement from the condemned (red label) to the neighbouring lands (blue label) category.

The medical officer normally had some kind of notebook which contained reports by street or court and indicated some distinctions between individual dwellings. At the Islington (Essex Road) Inquiry, Dr Tidy read from his notes regarding Cottage Place that 'the houses are dirty, damp and dilapidated. Floors close to the earth. Roofs very defective. Some of the houses have no back yards and are supplied with water butts in the houses in Popham St. 4–8 have small back yards, the rest none.'[7] These notes are evidently the product of simple visual inspection, and they were meant to serve as an aide-mémoire which would draw the medical officer's attention, and if necessary the attention of the commissioner, to aspects of buildings which could be linked to the conditions referred to in the Act. In any case of dispute 'worn out condition' or 'want of proper privy accommodation' would be examined by the commissioner, not through more detailed reports but by his own inspection. Thus at Essex Road he reported that three points had struck his attention 'in going over the area'.[8]

Eventually he would produce a judgement in such cases on which the confirming authority would act.

Remarks about the state of repair of dwellings were therefore couched in very general terms. Houses were 'dilapidated' or 'worn out' or 'in very bad condition'. For the purposes of Cross's Act, for a house to be condemned on the basis of repair or structural condition alone, this had to be fairly obvious. More usually, it was the 'closeness, narrowness and bad arrangement' of the property which was at issue. At places like Churchway and Coram Street courts lay not only behind front properties but also on lower ground where sites had been badly prepared. At Essex Road the site was held to be 'drainage ground' for the houses which were at higher level. Most of the houses had 'no cellerage or excavations for the foundations at all'. The subsoil was soaked, the walls of the houses were 'rotten and very damp indeed', and in the worst weather the whole site was said to be 'so ill drained as to be in fact under water'. This necessitated, it was claimed, 'the radical remedy of pulling down the houses, redraining the area, re-levelling the surface, and then erecting new buildings'.[9]

'Want of proper amenities' was a frequent product of confined space in association with multi-occupancy, and water supply, dustbins and privies were usually located in cellars or small backyards. In this respect the internal arrangement of houses sometimes posed special problems. Quite a number at Whitechapel (Rosemary Lane) were condemned because the tenants in the upper rooms had to pass through the ground floor rooms, through which the filth and slops in the upper rooms were carried, to get to the privies and water supply.[10] In some 10 per cent of the houses in the Boundary Street area 'the only access to the closet and ashbin' was gained by 'descending some rickety steps, then passing through a cellar without either light, ventilation or paving, and up some more steps into the yard'. Some of the cellars were only about 5 feet high and were used as 'receptacles for all sorts of abominable and useless refuse'. The inhabitants 'decline to use the yard and throw their slops and refuse into the streets'.[11]

The question of 'closeness, narrowness and bad arrangement' was posed in its most abstract form in relation to access to light and air. In cases such as Maypole Court, Rosemary Lane, the problem seemed evident. The ten three-storey houses, occupied by 86 people in 1871, lay around a court '80 feet long and nine feet wide'. 'The south end is blocked . . . by large buildings, and at the north end there is a high wall. The court is entered from Sun Court which is also a cul-de-sac, and Sun Court is entered from Upper East Smithfield by a narrow arched passage about 30 feet long, eight feet high, and less than three feet wide.'[12] Other less severe cases of this kind formed a frequent focus for owners' objections to the categorization of their houses. At the Whitecross Street Inquiry, the medical officer was asked whether he objected in principle to court property. He replied that it would depend on the width of the court, but declined to say how many courts there were of the same size in his district. The objector claimed 'to show that the effect of Dr Parry's argument is that we shall never know where to stop'. The Commissioner commented, 'that is the difficulty.'[13]

Although Cross believed that the 'plague spots' had been clearly mapped out 'by the aid of scientific discovery', approached from another direction the task of discriminating between different standards presented innumerable difficul-

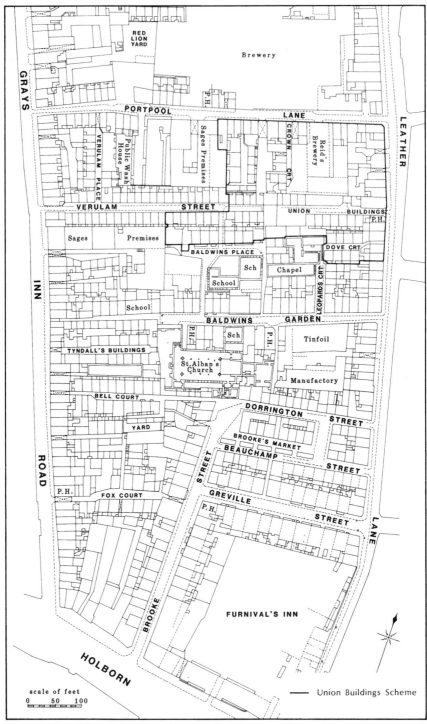

Figure 5.1 Holborn (1875) and Union Buildings (1899) clearance schemes. (Drawn from plan in MBW 1838/1.)

ties. Reporting on the grounds on which a house might be labelled 'unfit for human habitation', medical officers noted in 1891 that 'in all cases the question of degree comes in...[and]...renders precise definition impossible'.[14] The various grounds had to be balanced one against another, but any minimum standard had to be drawn at some point between finely divided properties. The main problem was, however, that no minimum standard was capable of being applied throughout London, due to the difficulties over rebuilding and rehousing. In the islands designated under Cross's Act, either the standard had to be set relatively high, so inviting comparison with much similar property outside the area, or it had to be set low, so leaving a large proportion of property within the area uncondemned.

The other general point of contention was the unalterable circumstance of 'closeness, narrowness and bad arrangement'. Owners objected that the condemnation of their houses bore no relation to their own efforts. Yet it was, of course, precisely because these circumstances were unalterable by the individual owner that they had been placed at the centre of the Act. This intended that houses otherwise not objectionable should be condemned where they were 'suffering from their neighbours'.[15] Nonetheless, the confirming authority, on the report of the Commissioner, could remove blocks of dwellings from the condemned category, and this practice seems to have grown during the period. Initially only blocks of some size were excluded, but in the 1899 schemes very small blocks or even individual houses were removed. At Union Buildings, Holborn, houses in Dove Court were moved to the blue category (Fig. 5.1). The Council complained, unsuccessfully, that although these premises had recently been rebuilt, they otherwise lay wholly within the condemned category 'being situated in a narrow courtyard approached under archways with insufficient yard space... and much too close to the houses in Union Buildings to the rear'.[16]

The area defined by the medical officer's representation (red label) could not be extended, and the 'neighbouring lands' consisted of additional property taken in order to make the scheme efficient for rehousing purposes. Co-ordination between those responsible for condemnation and rehousing was specifically ruled out, although the 'neighbouring lands' thus defined contained some property itself unfit. Nonetheless, the extent of blue labelling in a scheme does provide some indication of the manner in which the objectives of clearance and rehousing coincided as Cross's Act assumed they did. At Boundary Street, where precise calculations can be made from the compensation books, some 24 per cent of the residential property by value was blue labelled, and this category comprised 28 per cent of the whole property.[17] Among the 1899 schemes, the Council complained that, partly through Home Office decisions, at Roby Street, Garden Row, 'considerably more than half of the properties' were classed as neighbouring lands, while at Webber Row 'more than two thirds of the entire area' had been 'coloured blue'.[18]

The immediate effect of blue labelling was to provide an extra 10 per cent compensation. It also, however, guaranteed that market value compensation would be paid because such property was in effect declared fit. This categorization of 'fit' and 'unfit' took on an increasing importance as legislative provisions became more severe. The severity was in fact more apparent than real, but there remained a great deal of uncertainty both in regard to knowledge of how compensation procedures worked in general and as to how they might be applied in particular cases. The initial categorization of property could turn out to be decisive. By using its control over this procedure, the Home Office could thus to some extent protect owners against the legislative attempts to reduce compensation.

When the Board claimed in 1879 that it had 'generally been required to pay for the worst class of property as much as if there were no sanitary necessity for its destruction', its first suggestion was that the arbitrator should be instructed to consider 'whether it is possible by structural or other alterations to make the buildings fit for human habitation, and if so at what cost, and the amount of such cost should be deducted from the value of the property in its present state.'[19] The 1875 Act had already pointed in this direction. The 1879 Amendment Act stipulated that deductions should be made where property was in a state that made it 'a nuisance within the meaning of the Acts'. The 1890 Act added to this 'Or in a state of defective sanitation or not in reasonably good repair.' In addition, a rental could not be enhanced by overcrowding or use for illegal purposes.

The Board also stressed a second direction of legislative reform which offered a greater possibility of movement from market values. This was to introduce the concept of unfitness for human habitation into the compensation clauses. It was considered favourably by the 1882 Select Committee, but was noticeably omitted from the 1882 Act. The 1890 Act decreed, however, that where 'the house or premises are unfit, and not reasonably capable of being made fit for human habitation', compensation should be 'the value of the land and the material of the buildings thereon'. This seems a provision of wide application, since the main reason for which medical officers under Cross's Act were to condemn property as 'unfit' was its 'closeness, narrowness and bad arrangement'. These conditions meant in turn that it was 'not in the power of any owner to make such alterations as are necessary for the public health'. It would seem, therefore, that all the condemned buildings might be included within the scope of this provision.

The defence against the provisions contained many elements, each of which supported the other to form a strengthened whole. Discussion of land valuation and of the variety of property interests is required before this complex can be properly unravelled. Moreover, it has to be viewed in the context of political support or at least acquiescence from officials and legislators. A start can be made, however, by examining some of the elements which connect with points already made.

It will be evident that the provisions designed to provide good cause for deductions from market values raised similar problems to the procedures under Torrens's legislation. The complaints against the resulting rulings have a very similar ring. Thus the surveyor to the Board submitted to the arbitrator 'a detailed schedule of every item of repair that is required to put the premises into a habitable state'. He considered, however, that too often this was taken to involve 'a mere question of whitewashing and superficial alteration of premises'.[20] The Board's solicitor reported on the new clause in the 1879 Act that 'the evidence of the sanitary inspectors is that by a very small expenditure the premises can be taken out of the category referred to'.[21] The overcrowding provision was new, and apparently powerful, but it required evidence that individual rents had been increased by this cause. Yet one-room dwellings were not illegal, and let at the same price to a household of four members as to one of two.

Under Torrens's legislation, action had been limited by surveyors and magistrates. The Cross Act had removed the blockage at the expense of conceding market value compensation, but as the legislature attempted to withdraw from this position, it ran back into the old problems with arbitrators, surveyors, and juries. The effects on tenants could not now be used as an argument, but there were other defences. Landlords could still be thought responsible only for the remedies which it was reasonably within their capacity to effect, a factor again made more limiting by divided property interests. It meant that, except under extreme circumstances, the basic structural condition, situation, and arrangement of buildings could be left out of account. Moreover, arbitrators could interpret the provisions within a context of existing market arrangements, treating the executive authority on the same basis as any ordinary purchaser. Market valuations already took into account dilapidation and nuisances, but were primarily concerned with the extent to which these affected the future stream of rent income. In the case of slum property, this was in fact controlled by the manner in which magistrates interpreted the Torrens and public health legislation outside clearance areas. The Board was thus advised in 1884 that the arbitrator

> accepted and it is believed fully took into consideration the estimates of dilapidations which the Board by a competent builder was able to put before him. There is, however, nothing in the Act of 1882 which overrides the words 'fair market value', and though property . . . included in Artisan's schemes is of a very bad character, yet it is difficult to get over the fact that there is a market for such property.[22]

At Boundary Street, the valuer was instructed to consult a builder as to the probable cost of repairing condemned premises in twelve test cases. His valuations, however, as set out in detail in the compensation books, make no use of specific deductions from capitalized values except very occasionally in a token manner.[23] It was clearly his view that the amounts likely to be deducted under this heading could in any case be accommodated within the range of offers permissible under normal valuation procedures. Again, there was no general employment of the overcrowding or illegal use provisions. They were used most often against the proprietors of common lodging houses whose beds exceeded the licensed number.

In view of the preceding analysis all this is perhaps not surprising. Much more surprising is the failure to use the more powerful 'land and materials' provision which seemed to offer the possibility of escaping from market valuations altogether. At Boundary Street only bare land and unoccupied premises were treated in this way. In other schemes it was used to a limited extent for certain freehold property held 'in hand', but only where the resultant valuation was much the same as that which would have otherwise applied. What this meant was that compensation at Boundary Street and in other schemes was almost entirely assessed through ordinary market value procedures, together with the 10 per cent additional allowance for blue-labelled properties.

How then was the defeat of the land and materials provision for houses 'unfit and not reasonably capable of being made fit' achieved? The first essential step was that, whereas the legislation might seem to imply that property fell into this category simply by being condemned under Cross's Act, there was no actual stipulation to this effect. As a result, arbitrators were not prepared to accept that property ruled unfit by the medical officer was necessarily incapable of improvement, and the burden of proof was placed back again on the executive authority. This opened up once again the question of how far the situation of properties in relation to each other should be taken into account, and equally all the arguments over the arbitrary nature of any division between fit and unfit properties together with comparisons between properties inside and outside scheme areas. In effect the only acceptable proof was to obtain a closure order from the magistrates.

Under market valuations properties grade from one to another, and this is a great strength of the method. There is, however, no concept of the house 'unfit for human habitation', and dwellings can be supplied in any condition providing there is some demand. Attempts to raise the minimum standard through legislation were limited by the difficulties of arranging an alternative supply. When, however, such a supply could be arranged, as in a clearance and rehousing scheme, enforcement, in relation to compensation procedures at least, was limited by the insistence of arbitrators on the equitable treatment of proprietors inside and outside of scheme areas. The failure of the 'deduction from market values' method to push down slum property values while retaining a graduated approach led to the 'land and materials' concept, which accentuated all the existing problems of equitable treatment and added new ones through the separate assessment of land and buildings.

By retaining the graduated approach to property reflected in market values the arbitrators were at the same time protecting individual proprietors from inequitable treatment. This does not, of course, mean that the resulting outcome was 'fair', because the equity under consideration was solely that of one proprietor against another. The interests of the tenants, for example, did not enter into the account. Within this more limited context there was a certain logic in the arbitrators' approach, and, given the opportunity to do so, they were only prepared to withdraw protection from proprietors where there was evidence of a clear failure to carry out their responsibilites. The arbitrator Sir Henry Hunt gave as his view to the 1882 Select Committee that property condemned under schemes was

insanitary, but it is insanitary for various causes, insanitary because the houses are packed so close together, and the streets are very narrow . . . it would not be right to say that because . . . before the Building Act of 1855 was passed these houses had been erected with very narrow courts and passages, therefore they are insanitary, and the owner should be paid no greater sum than the value of the ground and the materials upon it.[24]

The fact that under Cross's legislation property could be found unfit for human habitation and incapable of improvement did not mean, therefore, according to this interpretation, that its owners should be deprived of market value compensation. The very ground by which property was included in the Act, the impossibility of improvement by individual proprietors, was a justification for that compensation. We do, indeed, return here to the logic of the original Act.

Compensation and land use

The essential context in which the Victorian slum developed was that increasingly obsolete built structures and environments were becoming increasingly more favourably situated in relation to sources of demand. This presented a certain paradox. On the one hand, increasing demand was a necessary precondition for the replacement of the slum, since new buildings would not replace the old until they were capable of yielding sufficient extra rent to finance the return on capital. On the other hand, property owners were continually responding to that demand in ways which increased the difficulty of achieving a satisfactory replacement. The creation of enhanced rentals through subdivision of property is an example, and infilling of remaining unbuilt areas is another.

Frequently these developments occurred together. It was a recurring theme of those describing the growth of slums that at some previous period, perhaps in the early part of the century, existing properties had been subdivided and garden ground and open space built over by cottages and other structures. At Bedfordbury front buildings were subdivided, like 37 Bedfordbury which had six rooms housing 33 people at the time of the representation, two rooms on each floor. Behind the front houses had been originally yard, garden or stable ground, some of which remained. Mostly, however, a series of courts had been developed approached through narrow tunnel entrances. The narrowest of these courts, Pipe Makers Alley, was only 3 feet wide and contained three houses of six rooms each. The larger Shelton Court had correspondingly larger buildings – six of the eight houses there had eight rooms each.[25] In these cases it was evident that as rentals were advanced, so making a more comprehensive redevelopment less likely, all the problems associated with constriction of space were increased.

Essentially the same problems arose in relation to business usage. The development of workshops in the narrow rear yards at Boundary Street was one of the contributing factors to the difficulties there. An example given by Harper shows the process at work at Albert Square, Stepney. In 1899 thirty-nine houses with rents of £35 per annum were bought for £22 000. The purchaser 'immediately proceeded to erect small workshops in the yards of the houses and to introduce alien tenants', so that by 1901 'rents had been increased to £65 and the value of the houses nearly doubled'.[26] The reference to alien tenants shows there were special circumstances in this case, but the capacity for constricting space while increasing rentals is evident; and if at the same time the front houses were further subdivided the residential densities might even be increased.

The piecemeal replacement of existing dwellings by larger structures was another process which might have deleterious consequences. In a case reported at the Churchway Inquiry, the vestry had ordered the demolition of six cottages under Part II of the 1890 Act. 'They were demolished and huge workshops three times the height and covering twice as much ground were erected in their place . . . these immense buildings block in the dwellings that face Drummond St. . . . the committee came to the conclusion that it had no power to deal with those buildings.'[27] The London Building Act (1894) set out to deal with the problem that legislation had 'gone out of its way to secure to all owners of old buildings the right to perpetuate and repeat insanitary conditions'. It was an important attempt at controlling the building of new structures in the midst of old, although it contained loopholes, notably in respect to commercial usages.[28] However, although such legislation might help to prevent the intensification of slum conditions, it needed to be complemented by the positive encouragement of more desirable reconstruction.

Cross's Act seemed at least to have the merit of recognising the effects of externalities and the need for some more comprehensive process of redevelopment. But these features arose more from the pragmatic resolution of perceived problems and suggested remedies than from any deep understanding of the nature of the slum. The slum as a 'mistake of the past' was over-emphasized, but there was little attention to the slum as a process of development. Concentration on the slum as a 'plague spot' meant that there was little understanding of its position relative to the overall geography of urban land use and land values. This was particularly true in relation to the intermixture of residential and commercial properties in slums, so that Cross's Act could omit the matter of compensation for trade disturbance altogether. Indeed, with the exception of public houses, the whole political debate on the slum in the Victorian period was confined to residential property invariably characterized in terms of 'wretched hovels', which was naturally expected to be of low value.

This misconception was soon apparent to the executive authorities. The very first representation to the Board at Holborn covered an area much infiltrated by business premises with surviving interior court populations (Fig. 5.1). It was realized that acquisition of business premises 'would add at least 50 per cent to the cost of purchase'. Of the 10 acres just over a quarter was 'occupied by Reid's brewery, Sage's factory, industrial dwellings and public washhouses, St Alban's church and national school, Elmslie and Simson's tinfoil manufactory and shops and premises in Leather Lane'.[29] At Shelton Street, the Board took pains to avoid business premises even of the most lowly kind. The result was, however, that 'good sites for workmen's dwellings are more or less spoilt by the buildings which are left . . . at the Drury Lane end the area is split up by the skittle alley being left in the centre of it . . . then there are the premises of Messrs. Jeakes which cut into another area. There is also Messrs. Corbens which divide the area into two parts.'[30] It was a sign of the reduced expectations of the administration that this site was accepted under protest, whereas that at Holborn had earlier been refused. Nonetheless, it had later to be extensively remodelled by the Council through the acquisition of business premises previously omitted.

In the early stages of the legislation, the executive authorities were inconvenienced by the application of the 92nd section of the Land Clauses Act, whereby taking part of a premises entailed an obligation to take the whole if the proprietor required it. This provision placed a considerable power in the hands of large businesses, being used for example by De La Rue at Whitecross Street to obtain a re-ordering of the scheme to their advantage. The position was that 'Messrs. De La Rue's premises are scattered over the entire area. There is a very large factory on the north side and also Star factory . . . if (they) enforced the entirety of their claim we might have £100 000 to deal with.'[31] Even after the 92nd clause was excluded, arbitrators could still award large damages for severance when part of a business was taken. This was the reason why the Board excluded Corbens, a carriage factory, from the Shelton Street area: 'In this case where a large trade is conducted on two sides of a narrow street the severance would almost amount to a complete destruction of the business.'[32]

The specific commitment of Cross's Act to rehousing ruled out any large-scale use of zoning or comprehensive urban renewal involving various land uses. It reflected Cross's intention that his bill should not become a 'town improvement' measure. In London this condition was still enforced even after the 1882 Act, notably in a judgement at Trafalgar Road, Greenwich. The Board's original plan involved two parts, one of which was to be developed for commercial purposes, the displaced population being rehoused in the second part. This plan was, however, rejected on the grounds that 'it was never intended that an artisans' dwelling improvement should have for one of its main objects the creation of an important commercial property at the cost of crowding the dwellings on the artisans' dwellings site'.[33]

Such a ruling had important repercussions, affecting the cost, incidence, and effects of clearance. As usual, however, there was a strong element of relative judgement involved in the application of the principle. There was no objection to the reinstatement of a displaced business in some other part of the scheme area, where this could be done satisfactorily. Moreover, marginal areas could also be developed for commercial purposes, and although of limited extent this was of some financial significance. The possibilities of recoupment through such development joined the avoidance of existing business premises as one of the considerations in the choice of sites. Later, Clare Market formed a notable exception in being developed entirely for commercial usages. In this respect, as in others, it was treated as part of the wider Kingsway project. It seems, however, to have opened the way for a more lenient treatment of the 1899 sites, several parts of which were entirely given over to business usage.

The relation of commercial premises to the slum formed part of the dispute over the extension of blue labelling at the turn of the century. Such premises could not be held unfit for human habitation, but they could be condemned as contributory to the conditions which made clearance necessary. This seems to have been the case at Boundary Street where red labelling was applied not only to stables and workshops but also to a silk factory, warehouses, and timber yards. But whereas some 60 per cent of business property at Boundary Street, by value, was red labelled, the effect of Home Office decisions in the 1899 schemes was to virtually exclude such property from condemnation.

It was claimed: 'It would cause hardship to condemn any commercial buildings which have been erected in due conformity to law and have been maintained in a sanitary condition.'[34] From this it appeared to the Council that the Home Office were now saying that 'all premises used for commercial purposes, however much they may be shut in or obstruct the light and air coming to dwelling houses, should be coloured blue'. They claimed that 'it not infrequently happens that the very worst conditions from a health point of view are found to arise from the too close juxtaposition of commercial buildings to residential property'.[35] In addition to shutting out air and light they might also form a nuisance. For example, ten stables at Webber Row had been coloured blue although the Council claimed that their filthy condition was one of the factors leading to the condemnation of the entire property.[36]

The value and type of business premises included in clearance can be established most accurately at Boundary Street where the detailed nature of the compensation books allows these interests to be separated from the residential property.[37] The total amount of compensation for business premises came to £94 660 or 36 per cent of the money spent in property purchase. The bulk of this money was for the purchase of buildings and went to the freeholders. This group, including ground landlords, received £47 920. Non-trading lessees received £12 837, usually for small business premises held on a larger lease with more substantial residential interests. Finally, there were 67 traders who received £33 909 for the value of their leases and trade compensation. Most of these were small men with three-year agreements, and 45 of them received less than £250 each. Below them were many more shopkeepers and workshop owners not included here since they had no longer-term property interest, and these were compensated along with the weekly tenants. In the case of condemned premises, trade compensation was one category which was standardized at half the rate normal in street clearances and other compulsory purchases. It amounted to one years purchase of net profits for passing trade, or half a years purchase where definite clients could be informed of a change of address.

Public house and beerhouse property was a special case and the only item consistently dealt with by a specialist outside valuer. The eleven premises at Boundary Street cost £32 410 to acquire, of which £15 874 went to the freeholders. Some £14 431 went to various breweries, principally Truman Hanbury, which held the freehold of The Ship public house and leased three others. The Ship was in turn leased until 1899 at £63 per annum, the lessee's books showing a net annual profit of £955 for which, with the lease, he received £2035. Truman Hanbury were in turn awarded £3290 for the premises, and seven years' purchase of their profit on the beer trade amounting to £2100, so that altogether The Ship cost the Council £7425. When the committee queried whether on the expiry of the lease 'the claimant could obtain the same rent on a 99 year lease' although the existing structure was 'old and worn out', the specialist valuer replied, 'The existence of the licence is an element of security which outweighs the condition of the building.' To put these figures into some kind of perspective, the Council bought the 30 acres for its suburban estate at Norbury in 1901 for £18 000.

The cost of these purchases was raised by the Council's determination not to allow the reinstatement of any licenced premises on their new estate. The size of the Boundary Street area did, however, allow this option to be invoked for some of the larger business premises. The biggest single units were Keeves, a 'large factory' for tin and japanned goods, and Vavasseur, Carter and Coleman's silk weaving factory. The latter favoured reinstatement since the weavers were 'all resident in the immediate neighbourhood'. Their premises were held on a 79 year lease at £20 ground rent dating from 1862 for which the freeholder received £630, and for the Vavasseur claim the Council paid £2750 for rebuilding. Keeves factory was freehold and valued at £8000 plus reinstatement.

The most striking group of trades, reflecting the industrial history of the area, was connected with timber and furniture making. Here the largest units were the timber merchants, three of which together received £4370 compensation, plus reinstatement of two of them. The cabinetmakers, upholsterers, wood turner and carver, music stool maker and so on were much smaller units, all receiving under £250 in compensation. Thomas Briggs, for example, was a fancy music cabinet maker in Jacob Street employing five men and supplying 'several large firms in Curtain Road and Tottenham Court Road'. His annual net profits were £200, and the claim was settled for £125. There were also a number of miscellaneous manufacturers, including a cardboard box maker (£1000) employing 30–50 hands, a machine maker (£500 plus reinstatement), brassfitter, waterproofer and bootmaker, marble mason and glass manufacturer.

Other trades were of the type that might be found in most slum districts. There was notably quite a large group concerned with stables, dairies, forage dealing, and barrow lending. Watkin Jenkins, the local head of the Welsh dairying empire in London, received £1048 for his business centred on a licenced shed for 23 cows in Mount Street. King, a forage dealer, had a warehouse 'of considerable extent' behind Old Nichol Street and accepted £550. Howard, a barrow lender, received £614. He had 'commodious premises' including 'two large sheds' in Charlotte Court with 320 hand barrows and 57 others for donkeys and ponies which he let to costermongers of the neighbourhood. Finally, there were the shopkeepers, general dealers, grocer and cheesemonger, bakers, fried fish dealer, ice cream and fruit vendor, all small men, and the missions and soup kitchens, including the large Mildmay mission and hospital in Turville Street. This was not, however, a lodging house area like Shelton Street or Mill Lane in Deptford.

Many of these trades would have been carried out in ordinary house premises, or in workshops erected at the rear. In other cases larger adaptions were required, which were gradually proceeding at the expense of residential property. Hurst's timber yard in Boundary Street and Old Nichol Street was 'old established' and occupied a 'large area'. There was a steam engine and twenty men and boys were employed. One lease had been taken out for 21 years in 1888 'in consideration of the lessee nearly rebuilding the four front houses and pulling down nine old houses in the rear'. Cording had a factory 'newly built' in Boundary Street employing three men and six to nine girls in waterproofing garments in a process using naptha. The firm had its headquarters in Regent Street, but their claim noted the cheapness of the Boundary Street premises and of the neighbourhood for labour. No objection was raised to the smell from the spirit.

As a bold project Boundary Street covered a large area and took less care to exclude business premises than many other schemes. By choosing areas further removed from central London it was possible to lower costs as the Board had done in the early 1880s. Boundary Street was, however, representative of conditions in the manufacturing belt of Victorian London. It has to be remem-

bered that in the calculations no allowance has been made for land given in
reinstatement, nor for the enhanced rents accruing to property from workshops
held on a weekly or monthly basis, a feature common at Boundary Street.
Closer in, as in the Union Buildings (Holborn) scheme, business premises
accounted for a larger proportion of property costs. The largest estate there, in
terms of compensation, was that of Watneys the brewers, for which freehold
£33 000 was paid together with another £10 728 for the leases on it and for trades.
It consisted only of two small shops, a public house, and 'two large warehouses'
each four storeys high in Portpool Lane, let to a manufacturer of artists'
materials and a wholesale chemist.[38]

Considering Union Buildings, the view advanced by the Council's valuer
regarding the Baltic Street area of Garden Row seems attractive:

> it is obvious that in a district of this character it pays better to utilize the land
> for warehouse and other kindred purposes than for dwellings, and there can
> be no doubt that the owners, as soon as expiry of leases permits, will
> develop the property for business purposes.[39]

However, even in a small area of 1.87 acres with many business premises, the
Union Buildings site still housed 907 persons in its remaining dwellings. It had
long been notorious when it formed part of the original Holborn scheme in
1875. It had been visited by the Prince of Wales during the Royal Commission
investigations as a prime example of slum conditions and considered by the
Council for clearance in May 1890. Many of the leases on the important estate of
Lord Leigh had in fact fallen in September 1890, but there had been no
redevelopment and the property had been relet on twenty-one-year terms.[40]

Apart from direct costs, commercial usages were also of major consequence
in any valuation of land. The normal rental method of assessment had at least the
merit of being based on definite information. The separate valuation of land,
abstracting from existing buildings and usages, had no such precision. The
arbitrator Sir Henry Hunt thus used a land-and-materials valuation in the case of
a large freehold property at Whitechapel (Rosemary Lane). He calculated on the
basis of 10s. a foot (6d. a foot and 20 years purchase) which amounted to £25 000.
He told the 1882 Committee that he had valued the land 'For what I thought he
could get for any purpose in the market.'[41]

There was a good deal of talk before the 1882 Committee of 'contingent
probabilities' in valuation. The Board's surveyor gave an example of the way in
which these came into consideration. At Whitecross Street houses lying some
distance from De La Rue's factory, and unlikely to be required for its extension
for some time, 'would not let perhaps at more than 6d. a foot to rebuild, but De
La Rue have been giving 1s. 6d. and 2s. for the acquisition of such properties as
are immediately contiguous to theirs, and therefore that would be cited as the
value'.[42]

It seems clear, however, that although the presence of some definite source
of demand might strengthen the case, potential commercial usage could be
successfully claimed on more general grounds. This was not affected by the
removal of the Cawley amendment to the 1875 Bill. When local authorities
attempted to sell cleared sites, the price was fixed by bids actually received.

In the compensation procedures the valuation of land was subject to no such constraint, and the assessment of future usage and value was inevitably hazardous.

Speculation on future land uses was, of course, not simply the product of compensation procedures. It has always been recognized as an integral part of the 'transition zone' concept. A modern characterization of this, viewed as a specifically historical phenomenon, is given by Griffin and Preston:

> The zone of mixed land use and blight presented a moving target...
> perpetually creating conditions of mixed land use on the outer periphery
> where business and industry penetrated residential areas, while on the inner
> side of the zone, conversion from residential to non-residential land use was
> becoming complete. In accordance with this process, the urban land
> market was presumed to adjust to this progression with the actual outward
> invasion being preceded by a wave of speculation. Speculation created
> artificial property values which in turn contributed to the physical and
> social deterioration of the threatened urban districts beyond the expanding
> urban core.[43]

Essential to this process is the manner in which individual parcels of land are brought to the market over a period of time. Had the public authorities been able to demolish all the property included in slum projects and schemes and bring it to the market altogether, it would have been evident that all of this land could not have been required for commercial purposes in any immediate future. Such a move was, of course, impossible because the existing occupants could not be displaced in such a wholesale manner. In a gradual operation potential commercial usage of some kind could easily be claimed for particular parcels, so that although the land was valued separately from the unfit buildings it was in a sense the existence of those buildings that protected the value of the land.

It should be emphasized that the hazardous value of land was not directly the cause of high compensation in Victorian slum clearance. As already noted, the land and materials provision was used very infrequently, and almost all the valuations were made on the basis of existing rentals. Even where the land was valued separately from buildings, the resulting assessment was often suspiciously similar to that which might have been achieved by the rental method. Thus, Hunt's valuation of 10s. per foot in the case mentioned at Whitechapel compares with the 9s. 6d. per foot given for land and buildings in the scheme as a whole. Except for bare land, the rental and land and materials valuations were always alternatives. Extensive valuation of land at a price higher than obtainable by the rental method was ruled out by political considerations. On the other hand, claims for potential commercial usage could be used to prevent any considerable lowering of compensation payments by means of the land and materials procedure.

Notes

1 Select Committee on Artisans' and Labourers' Dwellings, *PP* VII, 1881, 2331.

2 Sykes, J. 1901. The results of state, municipal and organized private action on the housing of the working classes. *Journal of the Royal Statistical Society*, vol. LXIV, p. 192.
3 Royal Commission on the Housing of the Working Classes. Report, *PP* XXX, 1885, pp. 22–4.
4 Land Inquiry Committee 1914. *The land*, vol. II: *urban*, p. 173. London: Hodder & Stoughton.
5 Artisans' and Labourers' Dwellings Improvement Act 1875, section 3.
6 MBW 1880, St Giles (Little Coram Street) Local Inquiry.
7 MBW 1879, Islington (Essex Road) Local Inquiry, p. 33.
8 ibid., p. 34.
9 ibid., *passim*; Select Committee, op. cit., pp. 569–711.
10 MBW 1881, Whitechapel (Rosemary Lane) Local Inquiry, p. 56.
11 Bethnal Green Vestry 1883. Medical Officer of Health Report.
12 Whitechapel Vestry 1874. Medical Officer of Health Report.
13 MBW 1878, St Lukes (Whitecross Street) Local Inquiry, pp. 97–9.
14 LCC Minutes HC 17 Feb. 1891 (13).
15 Select Committee, op. cit., p. 3217.
16 LCC HC Presented Papers 1898–1900, Bundle 39, 9 May 1900.
17 LCC HC Presented Papers, Boundary Street Claims (90–91).
18 LCC HC Presented Papers 1898–1900, Bundle 45, 16 May 1900; Bundle 46, 16 May 1900.
19 MBW 2411/7 Report no. 915.
20 Select Committee, op. cit., *PP* VII, 1882, p. 685.
21 MBW Minutes WGP 20 Oct. 1881 (47).
22 MBW 1373 Memo 8 Dec. 1884 no. 21.
23 LCC HC Presented Papers, Boundary Street Claims (90–91).
24 Select Committee, op. cit., *PP* VII, 1882, p. 330.
25 MBW 1877, St Martins in the Fields (Bedfordbury) Local Inquiry, pp. 19–20.
26 Royal Commission on Alien Immigration 1903, *PP* IX, pp. 115–25.
27 St Pancras Vestry 1893. Enquiry into the South St Pancras Scheme, p. 68.
28 LCC 1900. *The housing question in London 1855–1900*; Knowles, C. C. and Pitt, P. H. 1972. *The history of building regulations in London*, pp. 34–5. London: Architectural Press.
29 MBW 2411/7 Report no. 727.
30 MBW 1884, St Giles (Shelton Street) Local Inquiry, p. 425.
31 MBW 1878, St Lukes (Whitecross Street) Local Inquiry, p. 181.
32 MBW 1884, op. cit., 8–9.
33 PRO HO 45 10198/B 31375, pp. 40–2.
34 LCC HC Presented Papers 1898–1900, Bundle 39, 25 May 1900.
35 ibid., 9 May 1900.
36 ibid., Bundle 46, 16 May 1900.
37 LCC HC 'Presented Papers, Boundary Street Claims (90–91). The claims are arranged in alphabetical order.
38 ibid., Union Buildings (Holborn) Claims (88).
39 LCC HSG/GEN/2/3, section VII, 12 July 1899.
40 LCC Union Buildings Claims, op. cit., HSG/GEN/2/3, section VIII, 1 Dec. 1890.
41 Select Committee, op. cit., *PP* VII, 1882, p. 227.
42 ibid., p. 595.
43 Griffen, D. W. and Preston D. E. 1966. A re-statement of the transition zone concept. *Annals of the Association of American Geographers*, vol. 56, p. 341.

6 Compensation and ownership

Property interests

An American author has recently complained that 'too frequently a stereotyped monolithic view of landlords is espoused by those interested in oversimplified housing solutions for the city. Whether the stereotype is that of a multitude of small owners . . . or its inverse, a tightly knit group of exploiters, usually depends upon the prejudices rather than the knowledge of the commentators.'[1] This seems an appropriate comment on much Victorian discussion, which was additionally enlivened by the presence of a leasehold system, and by a predilection for building up arguments from an initial assignment of moral responsibilities. Beames wanted to begin his inquiry into the rookeries 'by asking under what landlords such traffic exists. Some one must be much to blame'.[2]

In any political battle the ground landlords carried the largest guns. After hearing Lord Shaftesbury and the Secretary of the Local Government Board, the Royal Commission of 1884–5 questioned Lord William Compton on the management of the Northampton estate in Clerkenwell. Later, they went on to interview a large variety of witnesses but none of them were leaseholders or middlemen. Kaufman in his book of 1907 is unsure 'how far head landlords are to be held responsible for this state of affairs', but he is quite sure that 'it should not be forgotten who really are the guilty persons, namely the slumlords'. It is the middleman 'who gets the benefit of high rents, of big sums for compensation'. Echoing widely expressed sentiments he maintains that

> the unscrupulous cupidity of jobbers, house knackers and property sweaters who farm the worst of the houses and stand between the freeholder and the occupier, . . . is responsible for much of the evil of overcrowding. These persons elude the most stringent of sanitary regulations and extract the most exorbitant rents, impoverishing the wretched tenants, who are entirely at their mercy . . . By their system of trafficking in slum property and botching it up with shoddy repairs, they perpetuate the slums.[3]

Another statement of these views is to be found in the *Quarterly Review* of 1884:

> By far the greater number of houses in the metropolis are let or sublet sometimes through five or six hands, the landlord getting a moderate ground rent, settled probably long ago when a ninety nine year lease was granted and independent altogether of the actual profit made by subletting. The extortionate rents, the scant accommodation and the neglect to repair, go to swell the gains of one or more of the middlemen or mesne lessees.[4]

Alongside the bloated middleman presiding over his empire, there was often an emphasis on the extreme division of property. Lessees in the slums were 'people who have saved a little money; people who have been in trade; they are not a nice class of person as a rule'.[5] For Brodrick, the poorer districts had been thrown into the hands of hundreds of investors holding a few houses apiece: 'The smallest alley is often held by a dozen different proprietors, among whom there is neither cooperation nor the ordinary sentiment of landlordism. As a rule, these properties are administered without the slightest concern for the class or quality of the tenants; not a farthing is expended upon them.'[6]

Opponents of ground landlords presented a very different view of the state of affairs. Their analysis of the development of the Somers estate provides an example. According to Harrison, 'In 1799 the estate was covered with good houses with long gardens and plenty of space, but in 1830 the Somers Trustees anxious to increase their ground rents built all over the gardens and created these wretched slums.'[7] According to another account these developments coincided with the arrival of the Midland railway terminus, 'but the downward course thus begun was hastened by the action of Lord Somers'. The leases were about to fall in and 'for the sake of a larger immediate return' Lord Somers 'gave extended leases at a considerable increase in rent . . . in this way he secured nearly £50 000 in premiums and an additional rent roll of £15 000 a year'. Half a million pounds had been paid to the Somers family since. 'Practically in this case the ground landlord's receipts represent the whole of the "unearned increment" for the house farmers and middlemen, who did not reckon for the harassing effects of modern sanitary legislation, have in many cases been so bled in expenses for repairs and alterations as to throw their leases up altogether.'[8]

Jephson believed that accounts of slum property ownership often stemmed from 'the efforts of the owners to repudiate the responsibilities for their predecessors infamous neglect and to shift the blame for the apalling state of affairs on the middleman'.[9] It was indeed around the predominant political position of aristocratic ground landlords that the debate was structured, so creating an emphasis on division of property interests on the one hand, or on large leased estates on the other. During the Boundary Street debate Fleming Williams 'held in his hand a list of the owners of the ground values in the deeper squalor of Bethnal Green', but he was able to make particular play with the fact that 'the owners of the ground values of some of the property were the Ecclesiastical Commissioners of England and Wales'.[10] The later history of the Council provided plenty of other examples, as in Somers Town, or on the Duchy of Cornwall estates in Lambeth. Reacting to these developments, the *Property Yearbook* thought it necessary to admonish those who supposed that ground rents were in the hands of a 'few Dukes or a handful of great public bodies'.[11]

Compensation necessarily involved a public examination of the structures of ownership. Where as in the present case, market valuation was a determinant of the sums awarded, information was required on rents, leases, and the flow of income from tenants to freeholders, leaseholders, and ground landlords. The Compensation Books which were compiled by the Council therefore allow a privileged view of property structures which is not normally available except through the accounts of special single estates. The Boundary Street books are particularly valuable in allowing a separation of non-residential property, and information from them will form the basis of the account that follows, supported by reference to other Council schemes or to the more summary Awards of Arbitrators which are available for the schemes of the Board. Except at Whitechapel (Rosemary Lane), however, the latter omit much property purchased by agreement.[12]

As already mentioned in the last chapter, compensation for residential property at Boundary Street was based almost entirely on the standard rental method. Beginning with the gross rental, the valuer attempted to estimate the net income accruing to the various property interests, and offered a number of years purchase upon it. Ground rents were treated as net income to their recipients and deducted from the gross rental of the leaseholder. A sum of 33 or 45 per cent was deducted from the gross rental for various expenses, and the composition of this will be discussed later. However, although the calculation of compensation was based on an estimated net rental, for comparison with auction values, which will be attempted later in the chapter, the years purchase of the gross rental is the key figure. This also has the advantage of being objective and reliable.

Table 6.1 sets out in summary form the types of interest in residential property at Boundary Street, together with statistics of gross income, purchase price, and years purchase. Points to notice at the outset are that freehold property held in hand accounted for 26 per cent of the whole gross rental, 14 per cent was let on long leases (over 21 years remaining), and 59 per cent on short leases. The latter was the least highly valued property in terms of years purchase, and long-lease property the highest. Ground rents accounted for 53 per cent of the residential property compensation and residential leaseholders for 20 per cent. It is noticeable that the leaseholders of residential property as a whole received only £31 000 in compensation, and short-term leaseholds only £16 500. The infamous middleman did not rate high on the scale of Boundary Street payments – the bulk of the money went to ground landlords and freeholders.

Table 6.1 Boundary Street: residential property interests (I).

Totals	Gross rental (£)	Purchase price (£)	Years purchase
in hand	4 544	43 021	9.47
long lease*	2 449	25 666	10.48
short lease	10 264	90 193	8.79
total	17 257	158 880	9.21
of which			
Ground rents			
long lease*	457	11 109	24.31
short lease	4 038	73 621	18.23
total	4 495	84 730	18.85
Leases			
long lease*	1 992	14 557	7.31
short lease	6 226	16 572	2.66
total	8 218	31 129	3.79

Source: LCC HC Presented Papers, compensation books.
*over 21 years unexpired.

The extent of property held in hand at Boundary Street was by no means exceptional in clearance schemes. Such freeholds comprised 36 per cent of all property by value at Rosemary Lane and about 28 per cent at Clare Market. Later, reporting on the 1899 schemes, the Council valuer estimated that 'quite 70 per cent of business properties' had been owned by 'the freeholders letting directly to the occupiers'.[13] This, however, included an exceptional amount at Webber Row, Southwark, where nearly 90 per cent of the compensation went to the principal freeholders, Quallett's Trustees. Moreover, although property both large and small seems to have been deliberately kept in hand, in other cases this position arose fortuitously through leases having expired during the period of the scheme. The chances of this were high, since these areas contained so much short-lease property. The relatively low proportion of compensation payments given to leaseholders is, however, a reliable and general feature. It was 25 per cent at Rosemary Lane, 19 per cent at Union Buildings, and about 20 per cent at Clare Market. These figures include many business leases with accompanying compensation for trade disturbance.

Taking both residential and business property, there were 71 freeholders at Boundary Street, including certain copyholders of the manor of Stepney, whose current incumbent received £900 for manorial rights. At the bottom end were those whose compensation, at around £150, represented the price of a single house. Forty-one freeholders received under £1000 in compensation, and another 21 under £5000. Together, these people held just over one-third of the total value of freehold property. The other eight freeholders took the rest, and four holdings, valued at over £10 000 each, accounted for nearly half the total. The two largest interests were those of Gwatkin's Trustees valued at £35 000 and Woolley's Trustees at £27 000. Socially, however, the list of landholders was headed by Baronness Kinross of Stowe and the Ecclesiastical Commissioners, whose medium-sized estates, valued respectively at £8400 and £6260, represented the twin aristocratic and corporate pinnacles of the English landowning structure. The size distribution of both freehold and leasehold residential property interests at Boundary Street is set out in Table 6.2.

The estates of the Baronness and of the Commissioners were managed in a broadly similar fashion. Both had some land in hand, but this came from the expiry of leases. The bulk of the Commissioners' residential property was let in one large lease to Street for 80 years from 1856, and that of the Baronness was let to Goward for 80 years from 1861. These arrangements meant that the ground landlords were well removed from any dealings with their property, but the price for this was a low income from ground rents of only £84 and £74 per annum respectively. By contrast, Goward's gross rental income, after payment of ground rent, was £607 per annum and Street's £631, helping him to an address well away from Boundary Street at Sydenham in south London. These were by far the most lucrative residential leaseholds at Boundary Street, and neither man held any other property there. Goward received £5175 in compensation for his interest and Street £5750, whereas the Baronness and the Commissioners received £4500 and £3260 respectively for their residential property.

These examples show that there were certainly cases where lessees took the bulk of the compensation payments for slum property as well as the larger share of income. This required, however, an original long lease with a good unexpired

Table 6.2 Boundary Street: residential property interests (II).

	Freeholders		Leaseholders	
	No.	Sum (£)	No.	Sum (£)
Compensation payments (£)				
over 10 000	2	57 871	—	—
5000–9999	3	16 658	2	10 925
2000–4999	10	30 969	3	9 903
1000–1999	9	11 630	2	2 125
500–999	14	10 543	7	4 922
under 500	15	2 994	21	4 111
total	53	130 665	35	31 986
Annual gross rental (£)				
over 2000	2	7 587	1	2 509
1000–1999	1	1 038	3	3 840
500–999	4	3 300	3	1 899
200–499	6	1 845	6	2 133
100–249	16	2 433	12	1 868
under 100	24	1 017	10	414
total	53	17 220	35	12 663

Note: the table does not include a small amount of residential property for which no gross rentals were calculated, usually because it was empty.

term. In most clearance areas, however, any original building leases usually had only a short period to run, whereas property tended to be relet on shorter leases. This practice was also found on some aristocratic estates. At Union Buildings, for example, property which produced a gross rental of £672 in 1902 was leased by Lord Leigh in 1890 for 21 years at £200 per annum ground rent. The share of ground rent as a proportion of gross rental income is notable here, but the two largest estates at Boundary Street managed to do even better.

The Woolley property was leased out in four large blocks, mostly on 21-year leases dating from 1883. The Gwatkin affairs were not so neatly arranged. Their property had 14 leaseholders for the residential part alone, with a variety of leases issued initially for terms of 11, 21 and 31 years, the latter dating from the early 1860s and the others from the early 1880s. From a gross rental income of £3315 per annum the Woolley Trustees managed to cream off £1675 in ground rents. Gwatkin's Trustees did not do quite so well, with figures of £4272 and £1617 respectively, but even so they obtained a substantial income. These were indeed very high figures, and this was reflected on capitalization for compensation purposes, the Woolley ground rents receiving only 12–15 years purchase whereas those of the Ecclesiastical Commissioners received 35 years. Even so, on the Woolley estate the ground landlord's compensation of £23 593 for residential property compared with £3307 for the lessees, and the respective figures on the Gwatkin estate were £30 115 and £5885.

The major lessees had large amounts of property which yielded a relatively high gross income but had low capital value. Nathan Chambers, the largest operator, provides an example. He held 33 houses in Jacob Street and Fournier Street on two leases from Woolley, and another 47 houses, mainly in Mount Street, on five leases from Gwatkin, together with six houses with shops and a block of model dwellings on the Turner Settlement estate in Boundary Street and Half Nichol Street. All these leases dated from the period 1880–3 and they were mostly of 21-year term, except the model dwellings. Leaving aside the latter, there was a gross rental income of £2208, ground rents of £1045, and a capitalized value for compensation purposes of £2047. The interest on the model dwellings, from an original 60-year lease at a ground rent of £40 per annum and current gross rental income of £302 per annum, was valued at £1553.

As already noted, the Council's valuer deducted sums ranging from 33 to 45 per cent from the gross rental to allow for outgoings. The latter figure usually applied where separate rents were received from subdivided properties, and the loss on empties and expenses of collection was greater. Some claimants sent in their own estimates; Street, for example, reckoned that from his annual rental income of £631, after payment of ground rent, he spent £90 on rates, £21 8s. on water, £4 on house tax, and £70 on repairs, totalling about 29 per cent of such income. Here rates, water, and taxes constituted a solid core of which rates were by far the most important element. Insurance was not allowed for by Street, but a small sum was mentioned for this by some other claimants. Beyond this solid core, however, lay a much more variable sum. Street allowed 11 per cent for repairs, but made no mention of empties or bad debts, although other claimants included empties with repairs. There was no mention of management expenses, including the lessee's labour or that of his agents.

Claimants had an interest in overestimating the net income on which compensation payments would be calculated. Only one put his outgoings as high as 36 per cent, and this included a large sum for insurance. On a 21-year lease improvements were often made to the property at the start and taken into account in the terms of the lease. Afterwards, the leaseholder's profit would depend on keeping down repairs, without running foul of the sanitary inspector, and pushing up rents without creating too many empties or bad debts. Here much would depend on the timing of the purchase in relation to the building cycle and other developments. Between 1880 and 1895 the Board of Trade estimated that rents in Bethnal Green increased by 20 per cent. This was well above the London mean of under 8 per cent, and to a large extent reflected the increased pressure on East End accommodation caused by the Jewish immigration.[14] Another important factor in determining profits was the method by which the original premiums had been financed, and on this there is no information.

It seems clear, nonetheless, that to take on a short lease on dilapidated property on which ground rents consumed 40 and even 50 per cent of the gross rental, and a large portion of the rest went in fixed and unavoidable costs, was a highly speculative venture. The low capital value of such leases meant, however, that large numbers of properties could be acquired at relatively low cost, and a larger operation probably gave economies of scale and spread the risk. As the ground landlord drew off a much larger proportion of the rental income, the lessee became more of a house manager, and there was no doubt truth in Shaw-Lefevre's contention that his was not a secure and realizable interest. It required the personal supervision of the holder all the time so that his position was more like that of a trader. Such lessees, he said, should be compensated as traders, receiving 3–4 years purchase of profits.[15]

In compensation valuations at Boundary Street the estimated net income of the lessee was put on a 7, 8 or 10 per cent table. This meant that Street's net income, with 45 years unexpired term, was at 7–7.5 per cent worth 12–13 years purchase. Chamber's lease of 19 houses in Jacob Street and Fournier Street, with an unexpired term of 12 years, was worth 7 years purchase on the 9 per cent table, that is £507. Expressed another way, this was 2.3 years purchase of his gross income from the property after deduction of ground rent. The average years purchase of such income at Boundary Street was 2.7 for leases of under 21 years unexpired term.

There were 38 smaller freeholders at Boundary Street owning residential property compensated at less than £2000. Only eight of these leased their property, and this invariably on long leases, four of them dating from the period 1807 to 1824. It did not mean, however, that such freeholders lived on the premises or nearby, keeping them under close supervision. Only one of the 38 gave an address in the scheme area, and only two others in Bethnal Green. The remainder had addresses in such places as Islington, Mile End, Stoke Newington, Poplar, and Walthamstow, with for the bigger holdings a larger sprinkling from further afield such as Barnet and Elstree in Hertfordshire and Bromley in Kent. Long leaseholders tended to resemble freeholders in their address patterns, whereas the short leaseholders showed a greater tendency to give local addresses. Of the main lessees on the Woolley and Gwatkin estates two gave addresses at Walthamstow and Dover. The others, however, represented a world of more local connections. Chambers and Quaintrell gave addresses in the scheme area. There was an estate agent from Dalston, auctioneers from Pentonville, and a timber merchant from Curtain Road, Hoxton. One lessee, the late John Wilson, had Chambers as one of his executors, together with Robert Chillingworth, a chenille manufacturer of Spital Square, who had an estate at Boundary Street valued at £12 500, mostly business property but with some housing which he kept in hand.

A figure such as Nathan Chambers seems comparable to the Mr Ball whose activities as a member of Clerkenwell vestry were highlighted by the Royal Commission. He was said to have more than 60 houses acquired during a period of 10–12 years.[16] Unfortunately, the absence of rate books for Bethnal Green in this period makes it impossible to check other property holdings which Boundary Street leaseholders may have possessed. It seems clear, however, that above the level of Ball and Chambers there was another tier of house farmer for whom there is no definite evidence at Boundary Street. The Select Committee and Royal Commission both made reference to the business of Thomas Flight who, it was claimed, from offices in Broad Street in the City controlled as many as 18 000 houses.[17] This was undoubtedly the summit of house farming activity, but mention was also made of a Mr Clements who was 'a second Mr Thomas Flight' and had a large amount of 'bad property in different parts of London'.[18]

Unlike the more local world of Chambers or Ball, a feature of Flight's activities was the enormous geographical range over which he operated. Flight died in 1877 and his operations were afterwards controlled by his wife Matilda. She received compensation payments in slum clearance schemes from Bowman's Buildings, Marylebone, in the west to Rosemary Lane, Whitechapel, in the east, and from St George the Martyr, Southwark, in the south to Coram Street, Churchway and High Street, Islington, in the north. All the sums

received were small, the largest being at Bedfordbury (£1050) and Great Wild
Street (£520). The Flights did not rely on close personal supervision or any local
political influence to maintain their profits. They reckoned that with economies
of scale and spread of risks they could cope with the kinds of demands likely to
arise from existing sanitary legislation. It was asserted to the Royal Commission
that 'medical officers generally throughout London' found 'that even though
the houses may be in bad condition, sometimes, any repairs that are called for by
the local authorities are immediately executed'.[19]

Although the rise of the house farmer seems a sure mark of the development
of the slum, it is more difficult to discover any type of trend among the
freeholders. The nature of freehold interests varied considerably from one area
to another, as one would expect from general descriptive accounts. In the
absence of longitudinal studies, one is left only with the general impression that
freeholders in the slums were still the same traditional landholding groups that
appear in other areas. There does not seem to have been any widespread selling
out either to the small man with less choice or to the specialist operator aiming
to assemble land for redevelopment, although the reported operations of a
syndicate at Warner Street, Holborn, may come into that category.[20] The
existence of the leasehold system sheltered the large freeholder from close
dealings with the slum, and the rising value of the land ensured the safety of
investments. Institutional factors may also have slowed down change. It is
noticeable that a very high proportion of freeholds in slum areas, and probably
in property generally, were in the hands of trustees. This reflected not only
historical attachments and the accumulation of land within families, but also the
fact that freeholds were one of the few investments considered safe enough for
trustees to engage in.

It is also noticeable that in no case at Boundary Street were there any
sublessees of residential property. In so far as a third person intervened between
the occupier and leaseholder, this took the form of an agent, or a weekly tenant
who took a whole house and sublet rooms. Subleasing of residential property
was also rare in other Council schemes, although everywhere it was practised in
respect of businesses. There were, indeed, more freeholders than leaseholders at
Boundary Street, and although in terms of compensation payments it was the
leaseholders who were predominantly the small men, this was not the case in
terms of the amount of property actually under their control (Table 6.2).

It is evident, therefore, that the extent of subdivision of property in the slums
was exaggerated in the political debate, just as this also overstressed the rôle of
aristocratic and corporate ground landlords. How does this effect the claim,
explicitly incorporated in Cross's Act, that landowners were prevented from
redeveloping their land by the presence of leaseholds and other divisions of
interest? The leasehold problem certainly appears much reduced by the related
circumstances that leasehold holdings were often relatively large and of relatively
low capital value. If house farmers could afford to buy up large amounts of
leasehold property in the slums, then so could ground landlords. Moreover,
there were numerous instances in which property that formed parts of slums
notorious since the 1840s was re-leased in quite large blocks at higher ground
rents. This practice continued right through the 1880s and afterwards. It is quite
clear that the house farmer did not deal solely in the 'fag ends' of leases

unbeknown to the ground landlord who, due to mistaken management in the past, had lost effective control over his property.

Against this there were clearly many instances in which the division of property interests adversely affected possibilities of redevelopment. This was so at Bedfordbury, the ground landlords most favoured example, where freeholds were held in very small lots with trustees much in evidence as well as division amongst several heirs. Leases had very varied unexpired terms, some long, although here again a large number of freeholds were in hand at the time of clearance. Another point to remember is that Cross's Act was specifically framed in terms of the needs of dwellings companies which favoured relatively large tracts of land for sanitary purposes and to reduce management expenses. The Peabody Trust claimed that without the Act they would have been unable to purchase five out of six of their 1879 sites.[21] These difficulties, however, arose partly from the narrow economic margin of the companies, which meant that there was little money to buy out interests unwilling to sell.

Finally, it needs to be recalled that business interests formed a major obstacle to large-scale land assembly for housing operations. This, however, merely points once again to the overriding importance of financial considerations in redevelopment. The circumstances in which business property could be purchased for residential redevelopment were clearly limited. Again, redevelopment for residential purposes might involve forestalling more profitable usages for some time to come. The private 'model dwellings' at Boundary Street do not suggest a very profitable return. Hall had a block of 35 tenements claimed to have been built in 1871 at a cost of £5689, including £4239 for buildings. In 1892 they returned a gross rental of £368 per annum and were valued for compensation purposes at £4400. The block of tenements leased to Chambers had a ground rent of £40 per annum, whereas each of the ordinary dwelling houses under his control was providing a ground rent of £10–12 per annum to Gwatkin or Woolley.

The primary control over replacement of slum housing was thus clearly that stated by the Council's valuer:

> It would not be likely that the houses would have been pulled down merely because of their insanitary condition, but only in the event of a redevelopment of the district offering increased financial results.[22]

Compensation and market values

Ultimately, the most important factor in compensation was the manner in which payments compared with market values. How far did payments in excess of the market contribute to high compensation, and how far did compensation procedures themselves affect the market in slum property? Ashworth, in the best account of the compensation question for this period, concluded that 'the assessment of compensation under the Torrens and Cross Acts was unreasonably profitable to property owners and a serious drag on improvement'. This was due partly to the provisions of the Acts, and partly to their interpretation in the quest for compensation under the Housing Acts became the province of the racketeer. In all schemes there were deliberate attempts to obtain excessive compensation and very many claims, especially in London, were completely fictitious. It was, indeed, alleged in the early years of the Cross Acts that only three per cent of the small claims were genuine. After all, it was always worth trying ... and if the claim was successful the success might be substantial. At any rate the prospect was sufficiently alluring for outsiders to try and buy up property as soon as they knew it had been scheduled for inclusion in an improvement scheme.[23]

These conclusions relate to the proceedings of the Select Committee and Royal Commission, which they accurately reflect. Whatever the truth of the matter, however, these inquiries relied heavily on the opinion of witnesses, and contain little evidence from which the modern observer can assess the extent to which abuse occurred. Moreover, there is a disturbing tendency, when sharp practice is alleged, to talk in terms of interests which received relatively small proportions of total compensation.

In the above quotation, for example, the words 'small claims' are of some importance. The arbitrator Rodwell's evidence to the Select Committee is much relied on here, and he pointed to the existence at Whitechapel (Goulston Street) of claims manufacturers who solicited work. But Rodwell was a man of explicit prejudices. The freehold and leasehold claims, he says, were 'comparatively a simple matter, but after all if you get reasonable people to deal with you get nearly what is right. What I do want to call attention to are the yearly, monthly and weekly tenants.'[24] However, the total amount he allotted to this group at Goulston Street was £13 000, whereas freeholders and leaseholders got £105 748. Again, much of the most specific evidence concerns shopkeepers, such as Newton of Boundary Street who claimed for a fictitious business, or the 'shrewd Irishman' Booth found at Shelton Street who had 'stepped in when it was known the houses were coming down, fitting up a small shop with a view to compensation'.[25] There was little evidence about trafficking in ground rents or freeholds, yet these attracted the largest amounts of compensation.

All this is not intended to deny that racketeering took place, only to question its contribution to high compensation payments. Moreover, racketeering can be taken to mean many things from optimistic claims to advanced purchases of property and various types of clear-cut fraud. It seems best, for present purposes, to try and bypass these problems and obtain some direct comparison between compensation values and market prices. Whereas for the Board's schemes only the amounts awarded are known, the Council's compensation books allow calculation of the years purchase of the gross rental. They can, therefore, be compared with the data on the purchase price and gross rental of property at London auction sales as recorded in the *Land and House Property Yearbook*, a source introduced into modern research by A. Offer.[26] The *Yearbook*, however, does not begin until 1892. Previously, sales are recorded in individual numbers of the *Estates Gazette*, but they are not classified by district until *c*. 1890. There are also some references to previous auction prices in the compensation books themselves.

Most of the evidence therefore comes from the 1890s, some time after the major official inquiries. Since the new provisions of the 1890 Act were little applied in practice, compensation procedures were broadly similar across the whole Victorian period. In the earliest schemes the arbitrator played the main rôle in determining valuations – at Rosemary Lane he dealt with all the cases. He issued a preliminary award, and then considered objections before making a final award. Then an appeal could be made to jury by either party where property was valued at over £500. After 1882 there was a simplification of procedures, with only one award and reduced scope for appeal to juries. By the time of the Boundary Street valuation procedures were well established, and all the larger claimants had professional advice. Nearly all cases were settled by agreement between the valuer and the claimant, although of course the valuer's offer was made in the light of what the arbitrator would be likely to award.

However, this relative similarity of procedure does not necessarily exclude substantial variations in the level of awards, for which there was considerable scope even within standard valuation practice. It seems likely, for instance, that even at the beginning of the legislation some reduction in valuation had been achieved compared with schemes in which substandard property had been acquired which had not been specifically labelled as 'unfit'. Thus the Board's surveyor reported that in street improvements the practice of the arbitrators had been 'to take all freehold property, whether good, bad or indifferent at 20 years purchase and all leasehold property . . . at 16 years purchase'. Here he is speaking of the net rentals. He claimed that under the 1875 Act 'no freeholder in the condemned area has been given more than 16 years purchase and the lease-holders have been computed at 14, 12 and 10'.[27] This suggests that there was some further reduction in levels of compensation in the Boundary Street scheme. Certainly it seem reasonable to suppose that Boundary Street valuations were amongst the most favourable from the official point of view.

Two methods have been used to attempt to relate compensation payments to auction values. A first approach was to seek previous auction sales of property included in the compensation books. This was done for Boundary Street, Clare Market, and the three main schemes of 1899. However, it yielded a total of only eight cases in which comparison could be reliably made. A second method is to compare Boundary Street valuations with all auction sales of house property in Bethnal Green made between 1892 and 1898. There was no major change in the rates of years purchase of property during that time. This yields, of course, a much larger number of cases, but raises the problem of the comparability of the properties concerned, particuarly in view of the absence of rate books.

Two cases at Boundary Street, reported in the compensation books, provide the strongest evidence for overcompensation. They both concerned members of the Wearing family. K. Wearing owned five unoccupied houses which were first valued on a land-and-materials basis at £633. It was then found that the property had been purchased in July 1890 for £350 and nothing had been since spent on it. The valuer's estimate was supported by Vigiers, the Peabody valuer, who thought the original purchase advantageous, and the claim was settled at £550. A. Wearing had the freehold of the Star beerhouse which was at first valued at £754. Here again it was discovered that the property had been bought in July 1890 for £375; it was claimed that the purchaser had made a good bargain and £550 was eventually accepted. As the Boundary Street area was represented in April 1890, these were almost certainly speculative purchases made with the intent of profiteering from a scheme. Neither of these cases involved residential property valued by the normal rental method, but three freeholds in New Church Street, Clare Market, let to weekly tenants were valued at £1148, whereas, apparently unknown to the valuer, they had been sold at auction shortly before for £900.

In other cases the valuation and auction price correspond more closely. At 14 Holles Street, Clare Market, a freehold property let on a 21-year lease from 1887 with a £38 ground rent was valued at £745, whereas it had been sold at auction in 1893 at £800. At 15 Stanhope Street, in the same area, the freehold of a house let on lease from 1886 at £80 per annum was valued at £1440 in 1897 whereas it had been sold at auction in 1892 for £1350. The freehold of 20 White Horse Lane, also at Clare Market, let to weekly tenants, was valued at £288 whereas in 1892 it had been auctioned for £300. The lease of 68A Leather Lane, Holborn, was valued at £243 whereas shortly before it had been bought at £265. At Boundary Street, W. Bentley was offered £874 for freeholds in Sherwood Place, but was able to show that he had bought the premises in 1887 for £950.

An interesting feature of the Bentley case is that when he brought forward his evidence of auction sale the valuer noted that this was 'strong evidence of market value' and that the arbitrator would not be likely to award much less. The sum of £912 was agreed. This put the claimant at an obvious advantage since he was not obliged to produce such a record unless it favoured him. A few claimants did, however, specifically mention previous purchase prices on their claim forms. The valuer clearly knew of the records of auction sales, and in one case at Boundary Street, the register of the *Estates Gazette* at Tokenhouse Yard was searched from 1870 to 1885 without success. It seems likely that the Wearing cases were revealed through their record in the *Estates Gazette* of 1890. After 1892 it would seem a simple method to check this record, but it does not appear to have been done systematically. There were four cases at Clare Market where the valuer appears to have known an auction price before making his valuation, but the other cases mentioned above escaped scrutiny.

In the compensation books the various interests in a particular property can be reassembled to give a total purchase price for that property. This cannot, however, be done from the record of auction sales. Ground rents of various kinds varied enormously in years purchase from under 15 to over 40. They benefited as the unexpired term of leases shortened, but years purchase was reduced for improved ground rents which consumed an abnormal proportion of rental income. For this reason, 69 per cent of Boundary Street ground rents by value were bought at under 20 years purchase, compared with 29 per cent by value in Bethnal Green as a whole. The scattergram (Fig. 6.1) shows that years purchase of leases was strongly determined by the unexpired term up to about 15 years, as one would expect. After that other factors enter in more. Whereas Boundary Street was distinguished by a predominance of short leases, in Bethnal Green, most of which was more recently built, the majority of leases had over 21 years to run. This again makes comparisons extremely difficult, but it would appear that Boundary Street valuations do not lie wholly outside a normal range, if the short lease property recorded in Bethnal Green was of roughly comparable quality.

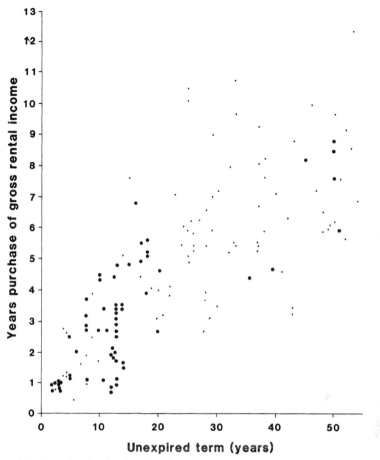

Figure 6.1 Leasehold valuations in Boundary Street, 1892–3, and Bethnal Green, 1892–9. (*Sources:* LCC HC Presented Papers, compensation books; *Land and house property yearbook.*)

Freehold sales form the strongest basis for comparative study. Table 6.3 shows the years purchase given for 122 lots of freehold property in Bethnal Green and 47 lots at Boundary Street and the reconstructed totals for the whole of the Boundary Street property, including recombined ground rents and leases. Most freehold property at Boundary Street was valued at 8–10 years purchase of the gross rental, whereas in Bethnal Green as a whole there was a wider range with a substantial proportion of sales at between 9 and 13 years purchase. Some 80 per cent of scheme property was bought at under 11 years purchase, compared with about 35 per cent at auction sales. Taking the condemned residential property only (Table 6.4), some 80 per cent of Boundary Street valuations were at under 10 years purchase, compared with about 19 per cent of the Bethnal Green freeholds. Boundary Street valuations were clearly in the lower range of property values in Bethnal Green. However, this is as it should

Table 6.3　Boundary Street, 1892–3, and Bethnal Green, 1892–9: freehold valuations.

Years purchase	Bethnal Green Value (£)	%	Boundary Street (in hand) Value (£)	%	Boundary Street (totals) Value (£)	%
3–3.9	510	0.4	—	—	—	—
4–4.9	—	—	—	—	—	—
5–5.9	—	—	496	1.2	496	0.3
6–6.9	665	0.5	—	—	2614	1.7
7–7.9	4675	3.6	3810	8.9	26234	16.5
8–8.9	6605	5.0	7446	17.3	40224	25.3
9–9.5	12955	9.9	12812	29.8	40734	25.6
10–10.9	20815	15.9	10319	24.0	14987	9.4
11–11.9	12420	9.5	8138	18.9	28729	18.1
12–12.9	14810	11.3	—	—	2274	1.4
13–13.9	14285	10.9	—	—	2466	1.6
14–14.9	3955	3.0	—	—	—	—
15–15.9	8475	6.5	—	—	—	—
16–16.9	4840	3.7	—	—	—	—
17–17.9	5390	4.1	—	—	—	—
18–18.9	5255	4.0	—	—	—	—
19–19.9	1025	0.8	—	—	—	—
20–20.9	2275	1.7	—	—	—	—
21–21.9	4205	3.2	—	—	—	—
22–22.9	3655	2.8	—	—	—	—
23–23.9	—	—	—	—	—	—
24–24.9	—	—	—	—	—	—
25–25.9	2800	2.1	—	—	—	—
26–26.1	1300	1.0	—	—	—	—
total	130915		43021		158758	

Sources: LCC HC Presented Papers, compensation books; *Land and house property yearbook.*

Note: value is auction price in Bethnal Green and compensation payment at Boundary Street. 'In hand' represents freeholds not subject to any lease. Boundary Street totals include leased properties where ground rent and leasehold payments have been recombined. In the Bethnal Green statistics extreme cases may represent errors in the reported classification of properties.

be, the question is whether the degree of difference between the two categories is sufficient.

Here there can be no definite conclusion. My assessment would be, as argued in the last chapter, that it would be unrealistic to suppose that all property in such a large and comparatively well-situated scheme had a lower market rating than any property in Bethnal Green outside the scheme. A degree of overlap is to be expected. Some support for this contention may also be found by examining the valuation of the condemned (red label) property within the scheme and that taken as neighbouring lands (Table 6.4). Leaving aside the additional 10 per cent given for compulsory purchase of blue residential property – a sum amounting to about £3900 at Boundary Street – the mean years purchase of the rental of the blue property was 10.25 and of the red 8.96. Each showed, however, a range of values, so that 23 per cent of the blue property was valued at less than 9 years purchase, and 11 per cent of the red at 11 years purchase or more. It does not seem likely that any demarcation of property on the basis of the criteria used in Cross's Act would at the same time represent a clear-cut line of division in property values.

Table 6.4 Boundary Street: valuation of condemned (red) residential property and neighbouring lands (blue).

Years purchase	Red property Value (£)	%	Blue property Value (£)	%
5–5.9	496	0.4	—	—
6–6.9	—	—	2614	7.3
7–7.9	26234	21.3	—	—
8–8.9	34459	28.0	5765	16.2
9–9.9	37558	30.5	3228	9.1
10–10.9	10466	8.5	4521	12.7
11–11.9	13188	10.7	15611	43.8
12–12.9	814	0.7	1460	4.1
13–13.9	—	—	2466	6.9

Source: LCC HC Presented Papers, compensation books.

Valuation is not a precise science, and a normal valuation may be more realistically viewed as a price band rather than as a single definite figure. Even so, scope still exists for coups at auction sales, particularly on the more special types of property. The evidence provided by the two methods described is by no means conclusive. It suggests, firstly, that only in very exceptional circumstances would property be valued for compensation at below market values. It provides also in the Wearing cases evidence of substantial coups. These were exceptional, however, as is shown by the other individual cases and by the general range of auction values. It seems likely that Boundary Street compensation was carried out within a perfectly normal range of valuation, although rather more generous than auction prices.

Some allowance must therefore be made for overcompensation compared with the market, but this ought not to be allowed to draw attention away from the two main reasons for high compensation in the Victorian period. These were, firstly, the high market value of central area property, both business and residential, and, secondly, the failure to drive down these values by reference to the politically powerful concept of the house 'unfit for human habitation'.

In the politics of slum clearance compensation, the struggle seems to be one between ground landlords and leaseholders, the former seeking protection by distancing themselves from the latter. This appears, too, in the actual provisions of the legislation with the separation of treatment between buildings and land. For although this clearly owes something to the fact that buildings might be described as 'unfit for human habitation' but not land, it also opened up a line of least resistance in the political sense. However, in the actual administration of the legislation a different alignment of forces appears in which the weaker elements of property successfully sheltered behind the stronger. This enabled politicians to quote the apparently stringent provisions of the legislation, even though the actual practice was quite different.

The differential effects which might arise from valuation on a land-and-materials basis can be seen in one of the rare cases at Boundary Street in which this method was used.[28] Some leasehold houses in Mead Street had become dilapidated and had been closed by magistrate's order, so that in this case there was some clear evidence of fault by the lessee. The houses had produced rents of

£80 per annum. Compensation to the leaseholder, who had 18 years unexpired, was calculated in terms of a notional annual value of the land, minus the annual ground rent of £12 12s., and yielded £36. The ground rent claim was valued at £346, or more than 27 years purchase, and was therefore fully protected.

As the enforcement of sanitary legislation became tightened, the very loose nature of the liability of leaseholders under ordinary law became an object of discussion. Thus 'the age and general condition of a house at the commencement of a tenancy' should be taken into consideration, and leaseholders should not be liable for defects 'caused by the natural operation of time and the elements upon a house the original construction of which was faulty'[29]. Yet this original construction was most important in the application of Cross's Act. We return here to the question of 'responsibility'. Although the 'responsibility of the owner for his property' could be used to drive down property valuations, it could also be a blunt instrument when used in relation to particular property interests, paradoxically treating leaseholders in a similar manner regardless of the way in which they had managed those elements of the condition of property which were realistically under their control.

Seen in this light, the action of arbitrators in not putting the land-and-materials provision into normal application, and thus protecting leaseholders, may again appear as a legitimate exercise of their function of ensuring equitable treatment. At the same time, however, officialdom readily accepted a generous interpretation of the market value of land which was the cornerstone on which the protection of freehold interests was built. If any substantial reduction was to be made in compensation payments, it was necessary to attack these stronger interests. This was part of the rationale of the Council's 'freeholder scheme' of 1900 which sought to draw attention away from the mere unfitness of property towards the rehousing obligations which this entailed. In this way, compensation for land could be assessed according to its value for working-class housing only.

The arguments in favour of this step were considerable for they bound together the destructive and constructive sides of slum clearance operations, whereas existing compensation practice related solely to the destructive element. A valuation of the land element of condemned residential property which took no account of rehousing needs passed this burden entirely from the landowner to the public authority. Arguably, the main purpose of the compensation payments in relation to buildings should have been to try and ensure their best possible maintenance prior to condemnation and while they were still required for use. It is very doubtful whether a land-and-materials basis of compensation could have produced that result.

Although the Council's efforts bore no immediate fruit in legislation, the 1919 Housing Act intended to ensure that condemned property was to be valued as a cleared site, and that to whatever extent the land was required for rehousing, the basis of value should be 'the value of the land as required for that purpose'.[30] In the debate Addison put the actual compensation cost of the Boundary Street scheme at £220 000 and its value under the terms of the new Act as £115 000.[31] It is not clear how either of these two figures were derived, but they illustrate the size of the effect the Act was intended to have. However, the LCC historians

Gibbon and Bell, writing in 1939, were able to sum up subsequent events as follows:

> The hard terms of 1919 were found to be too drastic for the public sense of justice . . . The difficulties were at first reduced by administrative practice; and in 1935 the law was radically altered. The full market value had to be paid for the cleared site whatever the purpose for which it was used. Full market value had to be paid also for any premises which were sanitary in themselves but unfit . . . because of their situation in a congested area.[32]

There has been a tendency to treat the Victorian experience regarding compensation as part of the prehistory of slum clearance, modern procedures beginning with the 1919 Act. An emphasis on racketeering conforms with this. However, although the 1919 Act put the public authorities in a much stronger bargaining position, the political struggle continued to be waged around all those considerations that had shaped the compensation question in the Victorian period.

Notes

1 G. Sternlieb, quoted in Stegman, M. A. 1972. *Housing investment in the inner city*, p. 7. Cambridge, Mass.: MIT Press.
2 Beames, T. 1852. *The rookeries of London*. Reprint 1970, p. 173. London: Cass.
3 Kaufman, M. 1907. *The housing of the working classes and the poor*. Reprint 1975, pp. 52–3, 17. Wakefield: E. P. Publishing.
4 Anon. 1884. The dwellings of the poor. *Quarterly Review*, vol. 157, p. 161.
5 Select Committee on Artisans' and Labourers' Dwellings, *PP* VII, 1881, p. 83.
6 Brodrick, W. 1882. The homes of the poor. *Fortnightly Review*, vol. XXXII, p. 422.
7 *Municipal Journal*, 2 Nov. 1893, p. 1637.
8 ibid., 13 April 1893, p. 173.
9 Jephson, E. 1907. *The sanitary evolution of London*, p. 297. London: T. Fisher Unwin.
10 Saunders, W. 1892. *History of the first London County Council 1889–91*, pp. 356–9. London: National Press Agency.
11 *Land and House Property Yearbook* 1892. Introduction. London: Estates Gazette.
12 LCC HC Presented Papers, Property Claims; MBW 1887, Awards of Arbitrators.
13 LCC HSG/GEN/2/3, section IV, no.60, 11 March 1908.
14 Smith, Sir H. Llewellyn. 1931–2. *New Survey of London Life and Labour*, vol. 1, p. 147. London: P. S. King.
15 Royal Commission on the Housing of the Working Classes, *PP* XXX, 1885, pp. 12647, 12652–3.
16 ibid., 617, 1141.
17 ibid., 4140.
18 Select Committee, op. cit., 2715.
19 Royal Commission, op. cit., 4485.
20 LCC Minutes HC 1 Oct. 1902 (51); 15 June 1904 (23).
21 Royal Commission, op. cit., 11710.
22 LCC HSG/GEN/2/3, section IV, no.60, 11 March 1908.
23 Ashworth, W. 1954. *The genesis of modern British town planning*, pp. 102–3. London: Routledge & Kegan Paul.
24 Select Committee, op. cit., 4894.
25 Booth, C. 1902–3 *Life and Labour of the People in London*, vol. II, p. 48. London: Macmillan.
26 Offer A. 1981. *Property and Politics 1870–1914*. Cambridge: Cambridge University Press.
27 Select Committee, op. cit., 513–4.

28 LCC HC Presented Papers, Boundary Street claims (90–91), claims of T. Watson and Woolranch Trustees.
29 Redman, J. 1896–7. Some legal incidents of tenancies of urban property. *Transactions of the Surveyors' Institute*, vol. XXIX, pp. 396–8.
30 *Hansard*, vol. 114, HC Deb., 3s., 7 April 1919, cols 1713–20.
31 ibid.
32 Gibbon, Sir G. and Bell, R. 1939. *History of the London county council 1889–1939*, p. 382. London: Macmillan.

Part III

TENANTS AND REHOUSING

7 Slum tenants and social policy

This chapter takes the form of a sandwich. It begins with the perception of the slum as a social entity and ends with a discussion of the manner in which slum clearance as a remedy fitted into more general patterns of sanitary and social reform. The middle of the sandwich is filled by two sections which deal with occupations, degrees of poverty, rents and overcrowding, mostly at Boundary Street. They bring together information which can now be gathered to characterize the social condition of the slum and, unlike the outer portions, are based on statistics presented in tabular form.

Model and reality

Barbara Wootton noticed that 'the assumption that there is a heavy concentration of the symptoms of social pathology at the bottom of the socio-economic scale has been the starting point for numerous social researches... Thus the founders of the C.O.S. could take for granted in their day that the poorest classes constituted a veritable cesspool of anti-social habits.'[1] This assumption was certainly present in the Booth inquiry, so that Duckworth's definition of the slum as an area which 'reflects the social condition of a poor, thriftless, irregularly employed and rough class of inhabitant'[2] was not a neutral linkage of poverty on the one hand and poor housing conditions on the other.

In Booth's scheme Class A was explicitly a moral class, but Class B was simply 'in chronic want', with earnings less than 18s. a week for a normal family. However, larger assumptions are evident in the care that Booth took to separate the casual labour of B from the 'irregularly employed' of C. The former class was not only poorer but also 'not one in which men are born to live and die, so much as a deposit of those who from mental, moral and physical reasons are incapable of better work'.[3] Beveridge, adopting the Booth scheme of things, sums up the distinction Booth was aiming at. Personal and industrial causes were prevalent in the distress of both Classes B and C. At the bottom, however, 'in these lowest types no doubt personal inefficiency appears to be the dominant cause of distress; the men would be unfit for anything else.' But B grades into C where 'it is the demand for casual labour that appears as the dominant factor, casual work is not so much chosen as tolerated unwillingly'.[4]

These distinctions were made in a search for remedies, a matter to which I shall return at the end of the chapter. The point here is that the characteristics ascribed to Classes A and B were also those ascribed to slum tenants. Indeed, the parallels in the approach to labour and housing conditions are striking. The problem was that 'the stern justice which competition deals out to the inefficient also has the effect of accentuating their inefficiency'.[5] Unemployment rendered those affected less suitable for steady work. Life in the slum rendered its occupants even less fit for tenancy of normal housing. In both cases they were liable to be sucked into that downward spiral of degradation which Victorian reformers so frequently invoked.

The industrial equivalent of the slum was the sweated trade. Here, says Churchill in the debate on the Trade Boards Bill (1909), the workers were recruited from 'the widow, the women folk of the poorest type of labourer, the broken, the weak, the struggling and the diseased'. The 'feebleness and ignorance' of these workers and their lack of organization 'rendered them an easy prey to the tyranny of bad masters and middlemen... themselves held in the grip of the same relentless forces'.[6] H. J. Tennant's remarks on the need to deal with the sweated trades were couched in language strikingly reminiscent of that applied to the slum: [These are] 'exceptionally unhealthy patches of the body politic where the development has been arrested in spite of the growth of the rest of the organism. It is to these morbid and diseased places – the industrial diptheritic spots – that we should apply the antitoxin of Trade Boards.' For, 'without drastic treatment they will continue as they have continued for three quarters of a century, feeding upon the national wealth, supplying recruits to our hospitals, asylums and workhouses, swelling the ranks of the unemployed and being the forcing grounds of the unemployable.'[7]

It is a commonplace of modern research that entities such as slums or sweated trades are defined and interpreted in public debate by outside observers. Modern writers point out that on investigation only a small proportion of the inhabitants of slums can usually be regarded as suffering from symptoms of social pathology. However Victorian slums may have differed from their modern counterparts, there is certainly a general problem in the degree to which contemporary debate seemed often to exclude any normality from the slum. The key factor here is the manner in which ideology guides experience. If the problem of perception is simply one of remoteness, whether social or spatial, then any misconceptions may be broken down by closer contact with reality. However, although this is certainly a source from which new concepts can be built up, it may not be immediately effective. Those who do learn in this way may not be able to communicate their knowledge to the vast bulk of the population who do not. More importantly, experience alone may not achieve an impact unless preconceptions have been weakened by doubts arising from the presentation of counter view. Otherwise experience may, by selection, simply confirm the existing viewpoint. It seems likely that general social conceptions become sharpened and reinforced as they become incorporated into particular political strategies, and it is at this stage also that spatial referents become more important. Once discussion begins to revolve insistently around a particular type of spatial entity, its distinctiveness and degree of internal uniformity almost invariably become exaggerated.

Arthur Morrison's view of Boundary Street in *A Child of the Jago* (1896) was itself cast in the form of an introduction to reality breaking down stereotypes formed in a world 'rotted through with sentiment'. The result, however, was to produce a deeper and stronger stereotype, emphasizing the slum as a place apart, its distinction from the norm and the dark and uniform blackness which existed within it. Morrison's Jago was entirely parasitical on the outside world, dependent on crime and charity. Within it only needy women normally engaged in useful occupations, and their need came from the absence of their husbands on 'ticket of leave', for 'cosh carrying was near to being the major industry of the Jago'.[8] True, Roper was a cabinetmaker, if unemployed, but then 'the Ropers were disliked as strangers because they furnished their own room... because Roper did not drink nor brawl, nor beat his wife, nor do anything all day but look for work'. They inevitably had to leave for 'were the lump once leavened by the advent of any denizen a little less base than the rest, were a native once persuaded into a spell of work and clean living, then must Father Sturt hasten to drive him from the Jago ere its influence suck him under for ever, leaving for his community none but the entirely vicious'.[9]

A very similar view was presented by John Reeves, School Attendance Officer for much of the Boundary Street area, and the man from whom Booth drew his detailed household data. Reeves's opinion was that 'the life of the people was chiefly occupied in deception and concealment'. They 'entertained an absolute dread of fresh air and cleanliness'. The picture he gives is uniformly bad except that 'the courts of the locality were even worse than the streets, pickpockets, burglars... dogstealers and pugilists, crime stained men and guilty women abounded'. Not surprisingly, he found the clearance scheme 'fully justified by the results... when we think of these slums, with every street, not to say every house, under the moral and physical taint, fostering the seed of disease and crime... was it not time for this great metropolis... to rise... and free herself from these dark vestments of the past.'[10]

Striking in its absence of nuance or attention to detail, there was nothing to distinguish Reeves's account from those of people wholly remote from Boundary Street. By 1903, however, a rather different view could be presented by the Chief Sanitary Inspector of Bethnal Green. He claimed that although at Boundary Street 'a large number were as described in "The Child of Jago" [sic]... I know some hundreds of families amongst them who were turned out... hard working, respectable decent people'. Perhaps this partly reflects the social status of a sanitary inspector. But he also believed that what is wanted is 'more – much more – house accommodation'.[11] The existence by this time of an alternative strategy surely encouraged his opinions regarding Boundary Street tenants.

In the case of property there was built into clearance procedures a mechanism which related individual holdings to each other and to holdings outside the slum. This was, of course, the market, a device which whatever its imperfections did nonetheless serve this purpose. It may be argued that by relating property to market values, an oversimplified model of the slum was exposed, giving some protection to property owners and indirectly to tenants. This may be seen as a justification for retaining a certain market-related element in compensation, especially in respect to buildings. Many tenants also received some form of compensation, although this was not specified in the 1875 Act. Hunt awarded £1 for each year of tenure up to a maximum of £15. Rodwell gave a removal cost for trades, otherwise token sums of £1 or £2.[12] The 1882 Select Committee recommended the Rodwell scale, but the practice of the Council was a compromise. It made payments of between 10s. and £5, varying according to length of residence, with larger payments for trade claims. Only in relation to trades, and in so far as they were linked to length of residence, did tenant compensation payments have any bearing on the nature of the slum. This, and their low scale in relation to total compensation, obviously ruled out any discriminatory function in relation to the selection of clearance sites.

Slum clearance procedures required certain statistics on the number of persons and rooms in designated areas, as these were related to the scale of rebuilding. Censuses were therefore carried out to produce these. Many other statistics were collected, such as those relating to compensation claims which will be used in later sections, but they were not subject to any contemporary analysis. Where this was attempted to some extent there was a clear political purpose, as at Clare Market where links between residence and workplace were examined in relation to the Council's claim that rehousing could be provided elsewhere. These investigations, therefore, constituted no independent stock-taking or inquiry into the nature of the slum, they derived simply from the governing strategy. As this strategy did not direct attention to the tenants, other than from a medical point of view, officials had little cause to find out much about them. Liddle, MOH for Whitechapel, did not 'go into minute detail as to what becomes of the people. I am engaged in improving the sanitary condition of the locality.' He nonetheless had 'not the slightest fear that inconvenience to any appreciable extent will be suffered by the people who will be displaced'.[13]

Another possible avenue for interaction between model and reality was through representation or protest. This worked best when it could be connected in some way with clearance procedures. Property owners, although not able effectively to oppose any schemes, nonetheless had the right as individuals to appear before the local inquiry and make representations regarding the grading of their property. Tenants had no such right, but occasionally groups were able to make a slight impact. Thus at Whitecross Street costermongers petitioned the Board of Works and were able to make representations at the local inquiry regarding provision of accommodation for barrows and donkeys.[14] In 1896 the Council received a petition from residents at Clare Market regarding the displacement of families to whom it was 'of the greatest importance that they should continue to reside in the district'.[15] In both cases some slight concession was obtained.

The best example of representation by slum tenants is, however, that provided by the inhabitants of the 'Italian Colony' faced with a possible scheme for the Warner Street district of Holborn. The area was said to be inhabited by 1170 persons, 'almost entirely Italians'. In 1901 the Council received a petition with 172 signatures stating that they were

> a very numerous and industrious people, united by ties of identity of race and country, and by our business occupations and pursuits whereby we are so connected that the breaking up and dispersion of the colony would be injurious in our respective vocations . . . we ought not to be deprived of the advantages of such associations . . . and removed from our place of worship close at hand.[16]

This petition no doubt reflects the special social structure of an 'ethnic village' organized around the padrone.[17] However, the successful warding off of wholesale clearance at Warner Street (at the expense of selective demolitions) owed much more to the fact that by this time the Council was divided on clearance policy. The valuer reported adversely on both acquisition costs and rehousing possibilities. For once the visibility of these people may have worked in their favour, for the effects of dispersion would have been more obvious, but the alternative was to rehouse 'aliens' at the expense of the ratepayers.

By the turn of the century this last factor may also have helped to deflect clearance from areas of Jewish occupation. Complaints were then being made that the new Boundary Street flats were 'full of aliens'.[18] Moreover, although Jews were often accused of being unclean and ignorant of sanitary ideas, this was not reflected in death rates. The Council MOH investigating Jewish areas found low mortality amongst 'people who were living in close courts and crowded alleys under conditions I was accustomed to find associated with very high death rates'. He attributed this to 'the better care the inhabitants take of themselves and their mode of life'.[19] But these considerations had not prevented the earlier clearance of Goulston Street, Whitechapel. Moreover, this clearance, and the representation of Bell Lane in 1877, was supported by the medical officer of the Jewish Board of Guardians.[20] Here social distance prevailed over ethnic links and drew the 'leaders' of the Jewish community into the prevalent paradigm of clearance and model dwelling construction.

The Irish stood in a different relationship to clearance schemes. They were not involved in the immigrant controversies of the last two decades of the century, but in the early and middle decades they occupied a central position in the debate on the slum and were by far the most numerous 'ethnic' group associated with slum clearance. They reflected more than most those aspects of lower-class life against which the social reform of the late nineteenth century was directed. The strong current of prejudice against the Irish must also have played some part in the designation of particular areas. Amongst the largest schemes, there were relatively few Irish-born heads at Whitecross Street or Boundary Street, but Rosemary Lane was overwhelmingly Irish and constituted the core of the Whitechapel area studied by Lees.[21] Although one may agree with Lees that such Irish districts possessed a 'closely-knit social life', they did not, however, have that more separate community structure of the Jews or Italians which in the case of pauperism, and to some extent crime, internalized these problems within their own group. Nor did they have any organized leadership taking part in clearance procedures.

Industry sections and social classes

In the census of Boundary Street taken for the purpose of the scheme, 1057 employed adults were enumerated, excluding 153 in common lodging houses. The main occupations listed were 149 labourers, 126 hawkers, 120 cabinet makers, 119 general dealers, 74 couch and chair makers and 74 shoemakers.[22] Fortunately, it is possible to go beyond this, since in East London Booth was able to fully implement his cross-classification of industry section and social class, details of which survive on a household basis in his notebooks.[23] These relate only to families with children of school age, 774 of which were recorded in the scheme area. The summary results of the cross-classification at Boundary

Table 7.1   Boundary Street: industry sections and social classes.

Sections divided into classes

Booth industry section	Booth social class (%)						Total	
	A	B	C	D	E	FGH	No.	%
labour (1–6)	7	61	9	12	11	—	181	23
artisan (7–12)	1	37	18	29	15	0	282	36
including Section 8	1	34	18	33	15	—	156	20
other wage (13–18)	—	24	6	47	24	—	17	2
home/manuf. (19–21)	—	15	19	19	38	9	47	6
dealers (22–27)	3	39	14	21	16	8	177	23
salaried (28–30)	—	—	—	50	—	50	2	0
female (33–38)	3	78	4	9	4	2	68	9

Classes divided into sections

Booth class	Booth industry section (%)								Total	
	1–6	7–12	8	13–18	19–21	22–27	28–30	33–38	No.	%
A	55	14	5	—	—	23	—	9	22	3
B	32	30	15	1	2	20	—	15	349	45
C	16	48	27	1	9	23	—	3	104	13
D	13	50	31	5	6	23	1	4	164	21
E	18	36	20	4	16	25	—	3	114	15
FGH	—	5	—	—	19	67	5	5	21	3

Source: Booth notebooks, British Library of Political and Economic Science.

Note: Section 8 is wood and furniture trades, and this is also included in the figures for the artisan group. Home/manuf. is home industries and manufacturing employers. Dealers include shop-keepers. For other sections see text.

Street are given in Table 7.1 and may be compared with the statistics given by Booth for Bethnal Green.[24]

One of Booth's main findings was that each industry section offered a relatively wide range of conditions of life. Labourers, for example, ranged from Class A to Class F, overlapping with the distribution of artisans, although the latter had a distinctly higher average condition of life. In part, of course, this simply reflects the manner in which these groups were defined. Artisans were wage-earning makers of goods, labourers included dock or building foremen, but excluded most transport, shop, or public service employees counted in the 'other wage' category. However, even closely defined sections offered a range of conditions with varying degrees of underemployment. The docks, at the lowest end of the industrial scale, provided in their pattern of permanent, preference, and casual labour simply a very exaggerated version of a more ubiquitous structure. For Beveridge there was always a reserve of labour whose scale reflected variability of demand and the state of organisation in each industry.[25]

Conditions at Boundary Street partly arose from the sectional composition of its employment. Proportionately, there were slightly fewer labourers than in Bethnal Green but twice as many dealers and female headed families. There were fewer artisans but more furniture workers than in Bethnal Green, where the proportions were 45 and 16 per cent respectively. There was a noticeable absence of 'other wage' earners and salaried employment, both more secure categories which together accounted for some 11 per cent of employment in Bethnal Green. More importantly, within each industry section Boundary Street people were disproportionately represented at the lower end of the range. Thus 68 per cent of labourers and 38 per cent of artisans fell into Classes A and B compared with 31 and 14 per cent respectively in Bethnal Green. Similarly, in the furniture trades 35 per cent of Boundary Street tenants were very poor compared with 16 per cent in Bethnal Green.

In all, 48 per cent of families were classified as very poor, with another 35 per cent below the poverty line and 17 per cent above it, compared with figures of 17, 27, and 55 per cent respectively in Bethnal Green. Labourers and artisans were the principal sections contributing to Class B, supported by dealers and female-headed households. Artisans made the principal contribution to Classes C, D, and E, together with some dealers and labourers. Those working on their own account made up most of the small number in Classes F, G and H. In classic fashion licensed victuallers were at the top of the social pyramid. Three of these were placed in Class H.

Booth's social classes were based on impression rather than statistics, and related to the condition of families, whereas the sectional classification was based on the occupation of the head alone. Class D, the regular poor, and implicitly Class C were to correspond roughly with earnings of 18s. to 21s. a week for an average-sized family. This was not a very wide range in view of the fact that boys, for example, might earn up to 4s. a week. Evidently, as Booth noted and Rowntree was later to emphasize, family size and life cycle stage could have a considerable influence on poverty rank. Class B, however, covered a wider income range, and at bottom included the destitute close to the workhouse. There was, for example, in the Booth notebooks at Boundary Street the crippled woman 'living in most wretched condition, no visible means of support'. The photo-tout who was 'wretchedly poor. Earnings about 8s. a week. Bears a good character but his clothes are so shabby no one will employ him.' Or again, the labourer with three children, one in fever hospital – 'a very pitiful case – the man has earned 3s. in six weeks, literally starving'. Individuals in such depths of poverty added to the reputation of the area.

Any assessment of social conditions must be distorted by the means taken to obtain the information. The statistics considered here can tell us nothing of behaviour, crime, or morality, as Booth discovered, and may sometimes be suspect in regard to employment. However, the general pattern does at least fit with the industrial character of the area revealed in the property compensation schedules, whereas it fails to fit with the wholly parasitic depiction of Morrison. Morrison's Jago, though, comprised only the southern part of the Boundary Street area, and Booth's data confirm that there was a social division here, as well as an expected contrast between the courts and the peripheral streets (Table 7.2). Even so, the contrast was more marked in terms of the grade of poverty than it was in industrial composition, and even in the southern streets artisans slightly outnumbered labourers. The Jago was certainly an area of deep poverty and marginal employment, but it cannot be cast outside the pale of normal economic life.

Table 7.2 Social divisions within Boundary Street.

Place	Total no.	Labour (1–6)	Artisans (7–12)	Booth labour section (%) Home/ Manuf. (19–21)	Dealers (22–27)	Female (33–38)
courts	125	27	29	3	30	10
periphery	103	17	36	10	30	4
northern streets	238	18	47	5	22	6
southern streets	308	29	32	7	19	12

Place	Total no.	A	B	Booth social class C	D	E	FGH
courts	125	4	61	12	18	6	—
periphery	103	2	33	17	2	33	13
northern streets	238	—	24	10	45	20	1
southern streets	308	5	59	15	11	8	2

Source: Booth notebooks, British Library of Political and Economic Science.

Notes
For labour sections see Table 7.1.
Periphery includes Boundary, Calvert, Church, and Mount Streets, and Virginia Road; northern streets are Newcastle, Jacob, Fournier, Mead, Christopher, and New Turville.

Variable social conditions were reflected in Booth's street colourations. Calvert Street, the outlet for the area into Shoreditch High Street, and conveniently on Ecclesiastical Commissioner's land, was coloured pink, indicating working-class comfort (Fig. 7.1). So were Church Street and Virginia Road, where relatively few buildings were taken, but Boundary Street was purple or mixed. All these streets lay on the border of the area. The northern streets and Turville Street were light blue – standard poverty – but most of the rest were dark blue – very poor. Old Nichol Street, Mount Street, Nichols Row, and some of the courts were coloured black. In Old Nichol Street this reflected the presence of 12 members of Class A out of 105 households. It was described in the notebooks as follows: 'an awful place, the worst in the division. The inhabitants are mostly of the lowest class and seem to lack all idea of cleanliness and decency... The children are rarely brought up to any kind of work and no doubt form the nucleus for future generations of thieves and other bad characters.'

Among Booth's school board blocks, Bethnal Green West with 28 701 people lay fourth in order of poverty with 59 per cent below the poverty line, and had the highest percentage of very poor – 33 per cent. The extent of its poverty was striking, but even within its core area at Boundary Street it was not uniform, so that there was not a very large contiguous area of black or dark blue streets. If the slum as an area which 'reflects the social condition of a poor, thriftless, irregularly employed and rough class of inhabitant' is taken to correspond with concentrations of Class B, then it took on a very fragmented pattern, to which the clearance sites of the Victorian period did not entirely correspond. Conversely, if the notion of slum be extended upwards to include streets

Figure 7.1 The Boundary Street area. (*Source*: LCC 1900. *The housing question in London 1855–1900*.)

characterized by Booth's Classes C and D, it would have encompassed areas too great for clearance to be contemplated.

It would appear that the fragmentary localization of extreme poverty is linked to the localization of employment. The Boundary Street area reflects – even if in very truncated form – the variety of classes within localized industry areas to which Vance drew attention.[26] According to Booth, 'The largest field for casual labour is at the Docks, indeed there is no other important field.'[27] Nowhere else was there such a large aggregate and such impersonal employment relationships. The dockside schemes, such as Rosemary Lane, contained a higher proportion of labourers, and no doubt a more uniform degree of poverty than Boundary Street. Other large clearance areas, however, such as Whitecross Street, resembled more the Boundary Street pattern, since they also lay within the Victorian manufacturing belt.

One employment localization that is particularly marked in relation to concentrations of Class B and clearance schemes is that of the street trades. The lists of street markets given by Mayhew or by Booth read like a roll call of clearance schemes: Whitecross Street, Clare Market, Chapel Street, Somers Town, Rosemary Lane, Leather Lane (Holborn), Aylesbury Street (Clerkenwell), New Cut (Lambeth). Such markets, although concentrated in the poorer districts of London, were also necessarily dispersed, and, as Mayhew remarks, 'The costermongers usually reside in the courts and alleys in the neighbourhood of the different street markets.' Even Boundary Street had a high proportion of dealers, and this was greater in some of the smaller schemes. Some, like Mayhew, were also inclined to stress cultural habits binding the costermonger to the slum: 'Home life has few attractions to a man whose life is a street life.' No doubt the high visibility of the street trader also strengthened this apparent link, and in their work as well as in their home this was one of the groups most affected by the more disciplinary approach of the late Victorian period.[28]

In London generally, Booth's dark blue streets probably contained a lower percentage of Classes A and B than at Boundary Street: his sample streets suggest 43 per cent. Taking these figures and those for the other street classifications, it can be shown that only some 23 per cent of Class B lived in streets labelled black or dark blue. The same percentage lived in light blue streets, but 36 per cent were in purple or mixed streets and 18 per cent in the pink streets that represented working-class comfort.[29] Such facts received little attention in a strategy aimed at concentrations of poverty. This, however, should not be regarded just as a cynical manoeuvre. The perception of social areas commonly exaggerates their degree of uniformity and distinctiveness as Cannadine has shown in the case of Edgbaston.[30] In the establishment of political strategies such stereotypes are built upon and enhanced. A typification of geographical concentration at various scales has always been an essential means of drawing attention to social problems, for whatever reason.

Rooms, rents, and overcrowding

Although the London cost of living survey of 1887 was confined to sample districts unrelated to any geographical structure, the Board of Trade inquiry of

1908 could place its findings within a scheme of zones and sectors. This loosely captured gradations in the nature of housing supply and rents, and the latter were found to diminish 'more or less regularly with distance from the City'. This inquiry was the first to give mean zonal rents for each tenement size. In the Central zone, for one-, two-, and three-room tenements in October 1905 these were respectively 4s. 6d., 7s., and 8s. 9d. In the Middle zone they declined to 3s. 9d, 6s., and 7s. 6d., and in outer London, largely beyond the County area, a three-room tenement could be had for 6s. 6d. These means were, however, produced from a considerable range: in the Central zone a one-room tenement might require from 3s. to 6s. and two rooms from 5s. 6d. to 8s. 6d.[31]

Although this range continued to reflect the varying advantages of location, it also incorporated variation in quality. Rooms were of varying sizes and existed in various states of maintenance. The position of rooms and access to amenities was particularly important in houses subdivided into small tenements. This was, of course, a particular feature of the Central zone. In boroughs lying wholly within this area, one-room tenements typically made up between one-fifth and a quarter of the tenement stock, and two-room tenements rather more than a quarter. In Middle Zone boroughs like Hackney, Poplar, and Greenwich, the figures were respectively 10 and 15 per cent. Within each zone tenements were larger in size in south London, and here rents were also relatively low. Even so, Pember Reeves, writing of Lambeth in 1914, believed that 'two rooms for 5s. 6d. are likely to be basement rooms or very small ground rooms through one of which perhaps all the other people in the house have to pass.'[32]

For most clearance schemes the overall pattern of tenement sizes and the mean rent per room are known.[33] Amongst the larger schemes (over 1000 displaced), the percentage of one-room tenements ranged from 95 per cent at Old Pye Street, Westminster, to 13 per cent at Webber Row, Southwark. It was about three-quarters in the West End clearances and at Rosemary Lane, and ranged between a half and three quarters in a large number of other schemes, including Boundary Street (51 per cent). The proportion fell to a third in schemes partly chosen to provide surplus rehousing. In the absence of general statistics for most of the period, the rents are difficult to compare with any meaningful standards, but in the larger schemes of c. 1900 the mean rent per room ranged from 2s. 7d. at Webber Row to 2s. 11d. at Aylesbury Place, compared with an average of 3s. 3d. in the Central zone in 1902.[34]

In all schemes there were considerable variations in the rent of rooms, and these can once again be illustrated at Boundary Street. The statistics are derived from data supplied by tenants claiming compensation. Five hundred and seventy cases were recorded, representing just under 40 per cent of the displaced households.[35] This is obviously not a representative sample, but it includes tenants of all categories, and some of the worst courts had large numbers of claimants. A further difficulty which must be emphasized is that some tenants claimed for rooms which they were subletting, particularly those who occupied the larger tenements. However, their subtenants could also claim. Compared with the scheme totals, the compensation schedules understate the proportion of one-room tenements (39 per cent as against 51 per cent) and overestimate the proportion with three rooms or more (27 per cent against 14 per cent), but the proportion in two rooms is the same (34 per cent.) Table 7.3 shows some

Table 7.3 Boundary Street: rooms and rents.

One-room tenements

Place	Percentage all tenants	Weekly rents (up to and including) (%)							Mean
		1s. 6d.	2s.	2s. 6d.	3s.	3s. 6d.	4s.	4s. +	
N and P	37	3	22	22	24	19	8	3	2.8s.
S and C	42	—	16	45	18	15	7	—	2.7s.
total	39	2	19	33	21	17	8	2	2.8s.

Two-room tenements

Place	Percentage all tenants	Weekly rents (up to and including) (%)							Mean
		3s.	3s. 6d.	4s.	4s. 6d.	5s.	5s. 6d.	6s.	
N and P	33	3	22	26	17	20	9	2	4.3s.
S and C	38	2	15	23	23	20	12	4	4.4s.
total	34	3	19	24	20	20	11	3	4.4s.

Three rooms or house

Place	Percentage all tenants	Weekly rents (up to and including) (%)							Mean
		5s.	6s.	7s.	8s.	9s.	10s.	10s. +	
N and P	30	7	16	14	21	13	14	14	8.0s.
S and C	19	28	19	15	21	13	—	4	6.7s.
total	27	15	17	15	21	13	9	10	7.5s.

Source: LCC HC Presented Papers, tenant compensation schedules.

Notes

North and periphery (N and P) included 260 tenements with mean rent 4.85s. South and Courts (S and C) included 239 tenements with mean rent 4.19s. In total 499 tenements had a mean rent of 4.54s.

For definition of places see Table 7.2.

distinction in tenement size between sub-areas of the scheme, but not as much as might be expected from the Booth classifications, except for the larger tenements.

The mean rent per room in the Boundary Street scheme was stated officially as 2s. 7d. The compensation schedule figures would seem to fall rather below that level. Rents were always proportionately higher in one-room tenements, and the mean of 2s. 9d. varied little from one part of the area to another. However, it varied considerably from house to house and for different rooms within houses. Thus, although about a half of one-room tenements were rented from 2s. 3d. to 3s. (and rather more in the South and Courts), one-fifth paid 2s. or less and 10 per cent 4s. or more. These variations were much greater than any mean interzonal differences, or sectoral differences within zones. The same degree of variation is found in two-room tenements, and there is a noticeable overlap between rents of the top range of one-room tenements and the bottom range of two rooms. Rents for three rooms or a house suggest that the latter consisted usually of three or four rooms.

It was not until 1911 that census data were published allowing calculation of the distribution of household sizes. Before that date, and since 1891, this could

Table 7.4   Household size distributions.

| Place | Percentage households in size group (persons) | | | | |
	1–2	3–4	5–6	7–8	9+
London (1911)	27	35	22	11	5
Bethnal Green (1911)	24	33	24	13	7
Bethnal Green (1901)*	26	35	23	11	4
Boundary Street (c.1892)	31	28	25	13	3
Boundary Street (1881)	27	25	30	14	3
Rosemary Lane (1871)	28	38	24	8	3

Notes

Data for Boundary Street, c.1892, were taken from LCC HC Presented Papers, tenant compensation schedules; data for Boundary Street, 1881, and Rosemary Lane, 1871, were taken from 10 per cent samples of Census Enumerators Schedules.

*One-to-four-room tenements only.

only be done for the inhabitants of 1–4-room tenements. That group had a larger proportion of small households than the whole (Table 7.4). Slum districts, as areas of very small tenement size, might be expected to take this trend further. Certainly they contained their full share of small households, and Rosemary Lane, where 70 per cent of tenements were of one room, had fewer large households than the norm. Boundary Street, with more two- and three-room tenements, had more larger households, and above the London average. Nonetheless, the impression of vast hordes of children which many observers gave arose from the density of habitation rather than the size of individual households. Perhaps the most striking feature is the relative normality of household size distribution in the slums, in view of their very abnormal distribution of tenement sizes.

These trends can be taken further at Boundary Street from the compensation data, but it must be remembered that there is some distortion due to subletting which, particularly in the larger tenements, exaggerates the amount of room space available per person. It is nevertheless clear that there was a relationship between tenement and household sizes (Table 7.5). In the whole area nearly half the occupants of one-room tenements were one- or two-person households, mostly the latter, and larger households were most frequently found in the larger tenements. In the poorest district of the South and Courts the proportion of larger households in smaller tenements increased, and throughout there were many large households living in one or two rooms. The result was a very high degree of overcrowding on the official scale, with two-room tenements, because of their larger households, having a degree of overcrowding (69 per cent) approaching that in single rooms (76 per cent). The Table shows, however, that very high proportions of overcrowding were to be found in one- and two-room tenements throughout Bethnal Green, a borough which stretched far east of the crowded district near the City in which Boundary Street was to be found. Thompson and others were certainly correct in believing that overcrowding, though worse in the slums and single rooms, was by no means confined to those categories, or uniform within them.

In the data available for all schemes there are two indices of overcrowding

Table 7.5   Boundary Street and Bethnal Green: households and rooms.

Boundary Street, c.1892: household distribution by tenement size

Tenement size	Percentage households in size group (persons)				
(rooms)	1–2	3–4	5–6	7–8	9+
one	48	31	16	5	—
two	23	28	33	15	2
three or house	18	25	26	21	10
total	31	28	25	13	3

Boundary Street, c.1892: population distribution by tenement and household size

Tenement size	Percentage population in household size					Persons	,Percentage
(rooms)	1–2	3–4	5–6	7–8	9+	per room	over-crowded
one	25	35	29	12	—	3.10	76
two	9	23	41	25	3	2.24	69
three or house	6	18	27	31	18	1.72	
total	13	25	33	23	7	2.20	

Bethnal Green, 1901: population distribution by tenement and household size

Tenement size	Percentage population in household size					Persons	Percentage
(rooms)	1–2	3–4	5–6	7–8	9+	per room	over-crowded
one	40	45	12	2	1	2.37	60
two	11	39	33	14	3	1.96	50
three or four	4	21	33	27	15	1.53	29
total	11	29	30	20	10	1.72	39

Source: LCC HC Presented Papers, Boundary Street tenant compensation schedules, Bethnal Green Census, 1901.

which allow Boundary Street to be placed on some comparative scale. The persons per room index is highly sensitive to the proportion of one-room tenements, and here Boundary Street, at 2.2, occupied a middle position among the 17 larger schemes. In its percentage of the population overcrowded, however, at 60 per cent it ranked sixth, being largely exceeded only by Shelton Street, Rosemary Lane, and Churchway. Not surprisingly, the schemes to the south of the river had the lowest degrees of overcrowding – below 40 per cent. In these figures, taken from special censuses, the overcrowding indices for Boundary Street correspond well with those produced from the compensation data in Table 7.5.

At Boundary Street two features were evident in the relationship of house-hold size and rent (Table 7.6). As households increased in size, so they tended to take more, or larger, rooms at increased rent. This feature stemmed from the changing composition of households as they increased in size, with extra adult and child earners. But not all larger households had this advantage, so that a small minority in the larger size groups remained in the lowest rental categories, and a much larger proportion did not increase consumption in proportion to need. Yet even in this legendary slum most households raised their rental

Table 7.6 Boundary Street: rents and household size. The table shows the percentage distribution of rents for households in each size category.

Rent	Household size (persons)				
	1–2	3–4	5–6	over 6	Total
under 3s.	46	22	11	4	24
3s. and under 4s.	21	27	23	15	22
4s. and under 5s.	14	21	26	17	20
5s. and under 6s.	10	17	18	19	15
6s. and under 8s.	5	5	6	27	9
8s. and under 10s.	3	6	10	11	7
over 10s.	2	2	5	8	4

Source: LCC HC Presented Papers, Boundary Street tenant compensation schedules.

payments in response to space needs. In part this may have been a compulsory increase in consumption, resulting from the application of overcrowding regulations in an area subject to special attention. However, Wohl is certainly much too sweeping when he says, 'the very concept of overcrowding and the desire for fresh air and ventilation were basically middle class; among the working classes there were many complaints about high rents and sanitary conditions but none concerning the lack of room space'.[36]

Sanitary and social reform

In principle, one objective method of demarcating the slum was through its association with high mortality and 'diseases indicating a generally low condition of health'. In practice this criterion could not normally be used to discriminate small blocks of property due to variation in population composition and the short period of years over which records were available. Even in the less exacting task of generally contrasting the record of the larger slum blocks with that of the districts in which they lay, considerable difficulties were encountered. The crude death rate statistics officially reported for the early schemes thus range from 54 per 1000 at Rosemary Lane over the period 1865–75 (London 23.6, district 26) to 25–27 per 1000 at Whitecross Street in 1873–5.[37] It seems clear that much of this contrast was the result of different recording practices by the medical officers concerned. At Rosemary Lane Liddle found that the death rate calculated from registration statistics was 38 per 1000 in 1875, but he had found 17 deaths from the area in hospitals and 51 in Whitechapel Union, which raised its death rate to 80 per 1000. By contrast Parry at Whitecross Street did not include such cases.[38]

In later years most medical officers included institutional deaths in their statistics. This did not, however, remove all the problems. The search for institutional deaths was confined to the represented areas, and a form of double counting may be involved. For though people from the slums were removed to hospital or workhouse, so others might be forced into the slum as death approached. Such statistical difficulties, of course, caused no real problem in getting schemes adopted. Indeed, during the 1880s and 1890s the crude death rates sited for scheme areas, though still varying, are generally higher than those for the 1870s schemes. At Garden Row, adjacent to Whitecross Street, the crude death rate was given as 40.7 (1895–9). It was 40.0 at Boundary Street (1886–9), 38.5 at Clare Market (1891–4), 30.6 at Webber Row (1896–9), and 48.6 at Union Buildings (1895–9). When Boscawen later pleaded 'we cannot hope to bring up a great imperial race in horrible slums', he reinforced his point by claiming, 'Notwithstanding all that has been done in nearly all our big cities we still have vast slum areas, places with appalling death rates of 40 or 50 per thousand.'[39] The Tabard Street area, which was his model London example, had a reported crude death rate of 36.8 in 1904–8 compared with rates of 18.2 and 14.9 per 1000 in Southwark and London respectively.

By the turn of the century, although diseases such as cholera and typhus had faded as concerns, there were still high death rates from phthysis and respiratory diseases. Infant mortality and rickets were high on the list of child-related problems. In all these cases, though other factors might be invoked as causes, there was still a prima facie linkage with insanitary conditions, and particularly with insufficient access of light and air. Thus, the LCC medical officer continued to stress the advantages of removing the slum, and indeed argued in 1899: 'The primary object of Part I of the Act is . . . to secure the removal from the midst of the community of houses which are unfit for habitation . . . Part I is not therefore in itself so much a Housing Act as an Act for removal of nuisances on a large scale.'[40] Moreover, the continued pressure of medical opinion regarding insanitary conditions could still connect with more general public sentiments such as those expressed by Dewsnup in his book on housing in 1907:

> They cannot be allowed to continue permanently in their insanitary and overcrowded conditions of living, frequently conducive of immorality, if for no other reason that they form a plague spot in the midst of the community . . . there is a heavy percentage of criminals amongst this class, and the condition of many others is the result of drunkenness, laziness and wasteful expenditure . . . it is to the public benefit that they should be driven out of their warrens into the light of day.[41]

Evidently this passage might have been written in the early Victorian period.

The great strength of the sanitary strategy which Simon had helped to formulate had indeed been the way in which it tied in the concerns of medical officers with wider aspects of public policy (Ch. 1). It strengthened the link between sanitary and moral reform, and offered a means of attacking social evils at selective geographical points. By operating on these points it would be possible to shore up the structure of society and prevent social deterioration. At the same time, by distinguishing between qualitative fitness and quantitative sufficiency of housing it offered the state a limited sphere of intervention which would not undermine the basic principles of political economy.

In making his analogy with the supply of food, Simon was perfectly aware of the medical damage caused by quantitative insufficiency as well as by bad food. In 1858 he presented a paper by Greenhow which drew particular attention to 'diseases from insufficient nourishment' and to the fact that 'children especially suffer from this cause; and many of their so-called scrofulous ailments are in fact mere starvation disorders, which a few weeks of better feeding can cure'.[42] What distinguished the qualitative case was not simply a medical priority but the ease with which a remedy could be fitted to current public policy.

Subsequent developments were, however, to confirm that the distinction between qualitative and quantitative could not be maintained. In the field of housing it was already compromised in Cross's Act despite Cross's own insistence that 'it was not the duty of the state to provide any class of citizens with any of the necessities of life'.[43] Later, the debate on one-room tenements and the increased attention given to overcrowding as an issue much reinforced the point, without bringing any clear solution. The standard remedy for an improved supply of commodities within the existing economic framework was larger-scale organization and increased efficiency. In London housing, however, the model dwellings companies and the municipal authorities had been successively thrown into the fray, but despite some advances they had not cleaned up the base of society.

One result was to encourage a retreat from intervention in housing, towards a purer market solution. Parsons thus stressed: 'Wage earners in London are not in a worse position in regards their house room than as regards their clothing or their nourishment.' He argued, with some justification: 'No item of family requirements lends itself more readily to separate consideration and treatment than the home. It is in view, it is always there; a defect in it whether in quantity or quality, offends or alarms a person of cultivated sanitary sense. There is a strong temptation to isolate this particular item from the group of family requirements, and to expect people to attain a standard of living as regards their house room which they do not attain in other respects.'[44]

In more progressive circles, interest was also turning to a more holistic approach. Hence Webb's famous call for the 'formulation and rigid enforcement in all spheres of social activity of a National Minimum below which the individual, whether he likes it or not, cannot . . . ever be allowed to fall'.[45] In particular, there was much interest in applying organization and efficiency in the sphere of employment rather than in the supply of particular key commodities. Indeed, the latter was seen to be dependent on the former. When lecturing on housing, Rowntree said, 'A wide extension of the policy of the Trade Boards Act, placing an increased number of trades under it and fixing a minimum wage for them is essential to true housing reform.'[46]

In its higher sphere, however, the effects expected of the Trade Board's Act bore some similarity to those of sanitary reform in its particular sphere of housing. In principle it would protect the better workman and employer against the unfair competition of the worst, so preventing a spiral of deterioration. The minimum wage would ensure this, but it would worsen the condition of those outside the pale. This was true too, of both the Beveridge scheme of 'decasualization' and Webb's plan, under which 'those whose labour in the judgement of the employers is not worth the national minimum . . . will be maintained by the community as they are indeed now'.[47] Instead of a more confused pattern of varying degrees of underemployment, with some possibility of movement between them, a more clear-cut division would be introduced between those who were to be protected from the abyss and those who would have to rely on the poor law. This was because, although the supply of key commodities might be guaranteed, in theory at least, by the minimum wage, there was to be no guarantee of the supply of employment at that wage.

The Trade Board's Act and the Beveridge and Webb plans can thus be regarded as a logical conclusion of a long development of a certain pattern of thought and action through the Victorian period. Minimum standards, conceived as a device for the prevention of social deterioration, had been pushed back from the particular to their most general point. Slum clearance can be placed within this development, of which it forms an important part. However, the suburban strategy which follows can not. By acting on housing alone, and simply substituting quantity rather than quality as the main target, in some ways it reverted to a more simple view. But this took place within an all-important shift of context, for the new strategy no longer involved the aim of shoring up standards by acting directly on conditions at the base of society. Instead, it was concerned with more powerful sections of the working class, and correspondingly, it involved a lessening of the disciplinary aspects of policy.

Cross's Act, I have stressed, had a complex rationale and was not a purely disciplinary measure. Some transfer of income from the ratepayers was envisaged in order to promote rehousing, although this was wrapped in all kinds of special pleading, and certainly not intended as a reward to the slum dweller. Once the scale of transfer turned out to be larger than expected, consensus was immediately broken. It was broken further when councils began to build themselves and had to set a notional housing value on the land rather than establish this by any kind of sale. Although the position of the municipalities was equated with that of the dwellings companies, it brought the question of subsidy more threateningly to the fore.

Cross's Act had, however, shown that some income transfer in the form of housing subsidies could be made palatable at a time when any direct redistribution could simply not be contemplated. It was less obviously a benefit to the recipient and could be directed to particular areas for particular purposes which could be claimed to be in the public interest. Liverpool Council's pushing over the thin line between subsidy on land and buildings at some time during the first decade of the new century also owed much to the stealth with which this operation could be accomplished, as well as a change of approach. It was possible politically only because Liverpool was not planning to act as a municipal trader.

The larger objectives of the London Progressives in this respect could only be sustained, in the political context of the period, by rejecting subsidies in favour of the alternative of land reform. It is important to see these as alternatives. In operations under Cross's legislation land reformers sought to replace a ratepayers' subsidy with a transfer of wealth from the landlords by means of reduced property compensation. Failure to achieve this end was a major factor in a switch of attention to the suburbs, where again capital gains which would otherwise have accrued to landlords in the process of urban growth were to be captured by municipalities, and indirectly by their tenants. Land policies would thus enable municipal action to have an important impact on the housing market without any subsidy from the ratepayer or taxpayer.

Gilbert, in his analysis of the origins of welfare, argues that the major innovatory feature was the granting of aid without penalty: 'Within this definition unsubsidized housing programmes are not social legislation but subsidized housing is.'[48] Here the Liverpool programme in its latter stages, and indeed all housing provided under Cross's Act, would seem to come within the bounds of the new welfare, whereas Part III provision would not. Such an interpretation would, however, be perverse, for if we look behind this definition to the larger spirit of Gilbert's analysis, it is the Part III strategy which fits best. For he says,

> By 1890 the old problem of pauperism had become a problem of poverty; and the essentially economic dilemma was political and social: what, it was asked, can the governors of the nation do to prevent the poor from using their franchise to overturn a society based on capitalist wealth? As it turned out, the defence against socialism was social legislation.[49]

The political strength of the working classes did not, however, lie with those of its members at the base of society. It lay not with the worker for whom poverty was endemic but with the one who 'feared the condition . . . knew that old age or accident to himself, a technological improvement . . . might throw him and his family onto the poor law'.[50] What he required was a safety net, and for this no plan that touched the poor law or carried a social stigma was acceptable, so that policies had to be built up in another direction. Similarly, Thompson's strategy began from the proposition that housing problems affected 'not the indigent poor alone but the great mass of working people of all grades'.[51] True, he did not propose to subsidize the housing for his artisans, but the benefits of municipal effort, of advance purchase of land and, indeed, of cheap travel, would initially go to them.

At the end of the Victorian period political intervention thus began to take on a more general form, in response to social problems which could no longer be associated solely with a pariah class. The battle over slum clearance fits within this wider context. No evidence can be offered here concerning the arguments which seek to relate the new welfare to the employers' interests, and to changes in their requirements for labour, but in any event it seems that the increasing political strength of the working classes, and reaction to it, was a vital ingredient.[52] What should be stressed also is that the new policies drew much of their immediate strength from the reactions to a long struggle in another direction, a struggle which we have followed through in the particular case of the slum. These reactions were not sufficient to confront directly and overturn the arguments which had supported previous strategy, and which often remained remarkably intact, but they were sufficient to cause a turning aside to new problems and new remedies.

Notes

1   Wootton, B. 1959. *Social science and social pathology*, p. 51. London: Allen & Unwin.
2   Duckworth, G. 1926. The making, prevention and unmaking of a slum. 33, *Journal of the Royal Institute of British Architects*, vol. 33, p. 328.
3   Booth, C. 1902–3. *Life and labour of the people in London*, vol. 1, p. 41. London: Macmillan.

4   Beveridge, W. 1909. *Unemployment: a problem of industry*. Reprint 1969, p. 144. New York: AMS Press.
5   ibid., p. 139.
6   Churchill, W. 1909. *Liberalism and the social question*, pp. 242–3, 240. London: Hodder & Stoughton.
7   *Hansard*, vol. 4, HC Deb., 5s., 28 April 1909, cols 351, 344.
8   Morrison, A. 1896. *A child of the Jago*, pp. 74–5. London: Methuen.
9   ibid., pp. 70, 150.
10   Reeves, J. 1913. *Recollections of a school attendance officer*, pp. 32, 59. London: Arthur Stockwell.
11   Royal Commission on Alien Immigration, *PP* IX, 1903, 6719, 6602.
12   Select Committee on Artisans' and Labourers' Dwellings, *PP* VII, 1882, 4495.
13   Select Committee, op. cit., *PP* VII, 1881, 300; Whitechapel Vestry 1878. Medical Officer of Health Report.
14   MBW Minutes WGP 22 Jan. 1877 (49); MBW 1878, Reports of Inquiries, Whitecross Street, 1877.
15   LCC Minutes HC 26 Feb. 1896 (14).
16   LCC HC Presented Papers 1901–1902, Bundle 66.
17   I am grateful to David Green for giving me a copy of his unpublished paper, The social economy of Little Italy in Victorian Holborn, Department of Geography, University of Cambridge (1977).
18   LCC Minutes 17 March 1903 (16).
19   Royal Commission on Alien Immigration, op. cit., 3963.
20   Whitechapel Vestry 1875–1877. Medical Officer of Health Reports. On this see White, J. 1980. *Rothschild Buildings*. London: Routledge & Kegan Paul.
21   Lees, L. H. 1979. *Exiles of Erin*. Manchester: Manchester University Press.
22   LCC 1900. *The housing question in London 1855–1900*, p. 91.
23   The notebooks are housed in the Booth Collection, British Library of Political and Economic Science.
24   Booth, C., op. cit., vol. 1, p. 74 (Table IX).
25   Beveridge W., op. cit., p. 13; more generally see Jones, G. S. 1971. *Outcast London*. London: Macmillan.
26   Vance, J. E. 1967. Housing the worker: determinative and contingent ties in nineteenth century Birmingham. *Economic Geography*, vol. 43, p. 95–127.
27   Booth, C., op. cit., vol. 1, p. 42.
28   Mayhew, H. 1861–2. *London labour and the London poor*. Reprint 1967, vol. 1, p. 11. London: Cass.
29   Booth, C., op. cit., vol. II, pp. 43, 229.
30   Cannadine, D. 1980. *Lords and landlords: the aristocracy and the towns 1774–1967*. Leicester: Leicester University Press.
31   Local Government Board. Report on the Condition of the Working Classes, *PP* LXIII, 1887; Board of Trade. Cost of Living of the Working Classes. *PP* CVII, 1908.
32   Pember-Reeves, M. S. 1914. *Round about a pound a week*, p. 35. London: Bell.
33   LCC 1900, op. cit., pp. 294–320; LCC London Statistics 1905–6, vol. XVI, pp. 119–126.
34   Royal Commission on London Traffic, *PP* XLI, 1906, 132.
35   Tenant compensation is recorded in LCC Presented Papers 1892–1896, Bundle A3, *passim*.
36   Wohl, A. S. 1977. *The eternal slum: housing and social policy in Victorian London*, p. 42. London: Edward Arnold.
37   LCC 1900, op. cit., *passim*.
38   MBW 1881. Reports of Inquiries. Whitechapel 1876, pp. 24–26; MBW 1878, Reports of Inquiries. Whitecross Street 1877, pp. 54–63.
39   *Hansard*, vol. 35, HC Deb., 5s., 15 March 1912, col. 1414.
40   Quoted in Jephson, E. 1907. *The sanitary evolution of London*, p. 365. London: T. Fisher Unwin.
41   Dewsnup, E. 1907. *The housing problem in England*. Economic Series no. 7. Manchester: Manchester University Publications.
42   Simon, Sir J. 1887. *Public health reports*, E. Seaton (ed.), vol. 1, p. 433. London: The Sanitary Institute.
43   *Hansard*, vol. 222, HC Deb., 3s., 8 February 1875, col. 97.
44   Parsons, J. 1903. *Housing by voluntary enterprise*, pp. 27–8. London: King.

45 Webb, S. 1911. The necessary basis of society. In *Fabian Tract* 159, p. 8. London: Fabian Society; see also Searle, G. 1971. *The quest for national efficiency: a study of British politics and political thought 1899–1914*. Oxford: Blackwell.
46 Rowntree, S. B. 1914. Manchester University Lectures on Housing, p. 29. Manchester: Manchester University Press.
47 Webb, S. op. cit., pp. 8–9; Beveridge, W., op. cit., p. 219.
48 Gilbert, B. 1966. *The evolution of national insurance in Great Britain: the origin of the welfare state*, p. 9. London: Michael Joseph.
49 ibid., p. 19.
50 ibid., p. 21.
51 Thompson, W. 1903. *The housing handbook*, p. 1. London: National Housing Reform Council.
52 On this see Hay, J. R. 1975. *The origins of the Liberal welfare reforms 1906–1914*. London: Macmillan.

8 Rehousing and dispersal

The new buildings which were erected by dwellings companies and the Council are probably the best-known part of Victorian slum clearance. They have been particularly studied by Tarn and Beattie, whose publications contain plans and photographs[1]. It is, however, the architectural aspects of buildings which have attracted most attention, whereas the purpose of this chapter is to link reconstruction and its effects to the nature of the whole clearance programme.

Setting the standard

When the Council first contemplated building at Hughes Fields, the Home Secretary hastened to point out that he could only sanction dwellings which would 'serve as a model for other districts'.[2] On the constructive side of the legislation, just as on the destructive one, there was a chain of responsibility in which the higher authorities were expected to maintain sanitary standards which others lower down were all too liable to relax. The assumption from which the whole legislation proceeded was that the slum had arisen because governments had failed to lay down a satisfactory framework of law within which commercial interests could operate. Society had to be based on those twin pillars of capitalist industry and responsible government. Here the rôle of government was a separate one, concerned with the sanitary and moral order. The minimum standards which it laid down were not, directly at least, intended as a contribution to working-class comfort, for that comfort had to be obtained within the commercial sphere of life. Similarly, it was for employers to pay and employees to merit the incomes necessary to support such minimum standards. Ultimately they were intended to come largely from greater effort and self-discipline on behalf of the poor themselves.

The new dwellings had to be beyond reproach also because they were intended as bastions of respectable life, which would spread their influence beyond, just as the old slums had contaminated their surrounds. Like the slum, too, they would have an effect on the training of the new generation. This was a major reason for opposing single-room flats, for as Sykes explained, 'Existence in a single room for a single man . . . or woman may be possible, but for families whose children will become future citizens and form the bulk of the nation it is utterly impossible without degradation and decadence.'[3]

Government departments aimed also to regulate the height and spacing of buildings to allow good access of light and air and to control densities. A maximum height of four storeys was generally recommended. Minimum room sizes were insisted upon for similar reasons. Balcony access was objected to, and the provision of staircases and water closets subject to special attention. In general terms a layout with good access was also required so that those stagnant conditions in which the back-street slum flourished should not be repeated. Widening of access to the Rosemary Lane site was thus seen as 'an essential condition to secure the occupation of the new buildings by respectable people'.[4] Such conditions varied little over the period, and of course they had important financial repercussions. Before examining these it is necessary to recall the vicissitudes through which building controls passed, producing frequent periods of compromise.

Matters were complicated initially by the Board of Work's belief that the new buildings were intended to cater for the persons displaced. It produced plans at Holborn and Whitechapel which envisaged mainly single-room accommodation 'of the plainest character' at prices similar to those currently prevailing.[5] This position was soon abandoned and at the Whitecross Street inquiry governmental opposition to single rooms was clearly stated by the commissioner.[6] Despite this, the Board continued to believe that its duty was to counter 'the undoubted tendency on the part of the companies ... to obtain the highest possible tenants at the highest possible rents'.[7] It envisaged retaining a strict control over the plotting of the land, the division of the tenements, and the nature of the buildings. This was the background against which the Waterlow and other companies refused to tender for the Board's land.

The result of the Peabody deal was that the Board gave up much of its control over the nature of the new buildings, in one case with unsatisfactory results. At St George the Martyr a plot in Gun Street was sold to Mr Snelling, but jerry-building was reported and later he went bankrupt. In 1885 the vestry complained that the buildings were in an 'insanitary state', and defending itself the Board stressed that the sale was made 'at a time when great pressure was applied ... by the Home Office in order to reduce the stringency of the conditions'.[8] More generally, five storeys now became the norm and the Home Office was forced to sanction six at Wild Street. Later both the Improved Industrial and Rothschild companies applied to build seven storeys, although they finally agreed to one less. It was in these circumstances that the East End Dwelling Company in 1884–5 obtained routine consent for its blocks of predominantly one-room tenements at Rosemary Lane and Goulston Street, Whitechapel.

The advent of the Council broke this course of events and brought renewed local authority control over the building programme. The Home Office saw an opportunity to return to its four-storey standard and to require minimum room sizes of 144 sq. ft for living rooms and 96 sq. ft for bedrooms. All this was reflected in the Council's report on building standards in December 1889, which also required more generous plotting of buildings, higher rooms and wider staircases.[9] Once again the dwellings companies baulked at the conditions laid down, notably at Cable Street where the Guinness Trust had begun negotiations.[10] This time it was the Council itself rather than the Peabody trustees which was to come to the rescue.

Another factor was that the first rebuilding took place on inherited sites the Council regarded as the Board's responsibility. Here a wide gap developed between the pre- and post-clearance use, with much lower densities afterwards. Even in the heart of London three-storey cottage tenements were used at Shelton Street in conjunction with a model lodging house. Further out, the use of such cottages at Hughes Fields and Trafalgar Road, Greenwich, was to form an interesting forerunner to the Council's later suburban policies. Once the 3 per cent resolution had been adopted in March 1893, however, an immediate break was applied to these tendencies. At Cable Street, where cottage tenements had once been considered, the Home Office was forced to acquiesce in five-storey tenement blocks. Five storeys also became the norm at Boundary Street, and shortly afterwards the cost-saving trend was continued by the construction on the second section of the estate of 'associated tenements' with closets or sculleries used in common.

Owen Fleming, in his account of the building of Boundary Street, says that 'towards the end of 1895 difficulties began slowly to increase'. Tenants demanded larger rooms and more self-contained tenements and 'their view found authoritative support in medical and government circles'.[11] Minimum room standards were correspondingly increased to 160 sq. ft for living rooms and 110 sq. ft for bedrooms and in many blocks exceeded these limits. In the meantime in 1898 the local Government Board effectively blocked plans by the East End Dwelling Company to build single-room tenements at Ann Street, Poplar, after the Council had agreed to abide by the Board's decision. It was only prepared to accept occupation of a room by a childless married couple, two girls, or two elderly persons of the same sex.[12]

Shortly afterwards building standards became a central part of the debate on future Council policy, and within that context a number of special circumstances contributed to pressure for building down. In the Clare Market scheme the Council was under special obligation to provide for Covent Garden porters and others on the Dukes Court and Russell Court sites as a quid pro quo for allowing Clare Market itself to be released for commercial purposes. The buildings were thus planned at relatively low rents, which brought Home Office objections to the reintroduction of balcony plans, the closeness of the proposed blocks, size of rooms, shape of bedrooms, thickness of floors, and the presence of only one staircase per block. In the end most of this was accepted. At Russell Court, however, a proposed 'family house', with common dining room, was abandoned when tenants objected to 'buildings conducted in the principle of common lodging houses'.[13]

The rise in costs at this time also brought pressure on building standards, particularly in low-rented areas. Although agreeing to new cottage designs at Trafalgar Road, the Secretary of State hoped 'the LCC do not propose adopting to any great extent cottages of so poor a type which he has approved . . . with great reluctance'.[14] Particularly cheap block dwellings were built at Ann Street, Poplar, and nearby at Cotton Street for Blackwell tunnel rehousing. In December 1902, however, the Council subcommittee on low-rental dwellings recommended that a rather higher standard of balcony dwellings with self-contained tenements should be built following a model used at Swan Lane, Rotherhithe.[15] The last phase of slum clearance rebuilding thus reverted to a more routine condition, if still rather austere compared with that at Boundary Street.

The cost of reconstruction

In a market economy development takes place only when it is profitable to the existing owners. In redevelopment of land already covered by improvements, the cost of land must at least equal the capitalized stream of current income unless it is clear that this cannot be sustained. In the circumstances of Victorian London it was rarely not sustainable even in the poorest property. Since redevelopment must also cover the cost of new buildings, it is evident that the process had to involve a very considerable increase in gross rental. For residential redevelopment this could be achieved only by higher densities or increased rent per room, or usually a combination of both. The land subsidy available in Cross's Act altered the quantification of these variables but left their fundamental import unchanged.

The Peabody purchase of 1879 set the marker for the price of land, valuing the five central sites of the Board at £10890 per acre and the Whitechapel site at £6377.[16] This amounted to only 3.25 years purchase of the pre-clearance gross rental from residential property alone. Although, as was claimed, the Peabody Trust acquired some of the prime sites of the Board on favourable terms, the overall value of its offer was not exceeded in later years. The six schemes by other companies that followed involved less central sites, the price per acre varying from £9000 at Goulston Street to £2000 at Wells Street. The average years purchase of the pre-clearance gross residential rental was 2.77. In the schemes of the Board left to be developed by the Council this figure fell to 1.21, but it recovered to 3.35 at Boundary Street, where £10710 per acre was paid, before falling back to 2.44 in the later Council schemes. These low years purchase figures may be compared with those quoted for freehold property in Table 6.3.[17]

In the passage of Cross's Act Waterlow had successfully amended the rehousing obligation so that schemes should provide for 'at least' the number of working-class persons displaced. He remarked, 'It was easy enough to arrange for the accommodation of twice or thrice as many persons to the statute acre as there were at present.'[18] In view of existing densities this was a very large claim. There were already some 600 persons per acre at Bedfordbury and Great Wild Street before clearance. Further out, however, densities were doubled on sites such as Bowman's Buildings by bringing the number of rooms per acre up to the levels prevailing in the centre. When the Council began to come to terms with its rent resolution, the effect was to immediately increase the number of storeys in the Boundary Street blocks from four to five. It was estimated that the cost of Streatley Buildings would be £11100 at four storeys and £12500 at five. However, the projected rental would allow expenditure of only £9000 on four storeys compared with £12200 on five.[19] Payments were made only for building plots, making no allowance for any extra street or other consequential space provision. There can be no doubt that on these terms the marginal revenue was greater than the marginal cost, despite the inconveniences of living on higher storeys without lifts. Moreover, the number of applications from companies to develop at six or seven storeys shows the commercial pressure for even higher buildings, and the significance of the struggle with the Home Office over building heights.

In the initial Peabody schemes, where only a slight overall increase in density was achieved, the cost of land amounted to £20 per post-clearance room. This was comparable to £23 at Boundary Street and £21 in later Council schemes. In contrast, by building on cheaper land at higher densities the initial six non-

Peabody schemes pushed the land cost per room down to £10.5, and it fell to £9.5 on the schemes of the Board which were completed by the Council where lower densities were compensated for by even cheaper land. The real cost of the Cross Act sites was, of course, much higher. In the initial Peabody purchases it amounted to £146 per post-clearance room, so that there was a subsidy of £126 per room, and the actual price paid by the company was only 13 per cent of the real cost. In the six post-Peabody schemes on cheaper land the subsidy per room fell to £86, but the price paid per room was only 11 per cent of the real cost. In the Board schemes completed by the Council the subsidy reached £127 per room, despite the suburban location of the sites, and the price paid was only 7 per cent of the real value. The subsidy was then much reduced at Boundary Street where £98 per room and 19 per cent of the real value was paid. Later council schemes saw this proportion fall to 12 per cent, but the high subsidy of £149 per room reflected the return of clearance to more central sites.

Despite the attention given to model dwellings companies, little research has been carried out on the manner in which they acquired sites or the financial calculations involved. It is known that on all the sites in London bought by the Peabody Trust between 1864 and 1885 the average land cost per room was £23. On the ten sites acquired before Cross's Act costs had been £8963 per acre or £21 per room. However, payments ranged as low as £3615 per acre and £7 per room for the Shadwell site to £20 691 and £34 respectively at Westminster. Indeed, on four sites land costs were over £30 per room.[20] Nevertheless the Guinness Trust acquired six sites in 1890–1902 at an average of £14 per room, ranging from £9 at Vauxhall Square, Lambeth, to £18 at Lever Street, Finsbury.[21] In its operations outside of Cross's Act, the Council paid £129 per room for the Reids Brewery site on which the Bourne Estate, Holborn, was built, and £161 per room on the land bought from the Duke of Bedford for Clare Market rehousing. However, it paid £21 per room at Millbank and £23 for the Caledonian Asylum land. It is clear that the last two purchases, apart from their less central location, were also arranged on special terms, and it seems that this must also have been the mechanism by which dwellings companies obtained cheap land. In one case at Chelsea the Guinness Trust was given the land on which to build. However, a great deal also depended on the careful selection of sites – some of which were cleared land – their layout in relation to streets and existing services, and their density of development.

A striking feature of the land subsidy in clearance schemes was that it usually exceeded the expenditure per room on the new buildings. The effect of the distinction between land subsidy and building subsidy, which was a very fine one particularly when Councils built themselves, was that much effort was spent in reducing costs on buildings which often achieved relatively small economies when set against the high and variable subsidy on the land. The building costs of the Peabody Trust on the Cross sites ranged from £82 at Coram Street to £92 at Whitechapel. The average building cost of all Peabody dwellings over the period 1864–5 was £86 17s. per room. The building costs of the Guinness Trust in the 1890s averaged £73 6s. per room, increasing at the end of the period to £83 9s. per room.[22] These figures may be compared with an average cost of building at Boundary Street of £91 11s. per room, a figure which, however, conceals substantial variations from block to block.

Figures quoted for building cost per room sometimes vary substantially, and the precise basis of calculation is often not stated. Nevertheless, it seems clear that costs per room in the initial Council projects were very high. This was still reflected in the first Boundary Street block, Streatley Buildings, which cost £107–116 per room. On Section B costs were reduced. Part of the reduction probably came from changes in design, allowing, for example, more flats per floor and per staircase. Self-contained flats could thus be built, as at Shiplake Buildings, for £72–82 per room. There was also in some buildings a departure from the self-contained principle, the architect presenting plans in 1893 whereby 'the estimate of £87 per room of the latest type of self-contained tenement might be reduced to £72 per room'.[23] This was achieved in Sonning Building, which had a private WC outside each flat and a shared scullery.

A Council report of 1896 comparing Boundary Street dwellings with those of the Guinness Trust in Stepney Green put the building cost per room of the Trust at £79 8s. compared with £71 16s. for the associated tenements on Section B of Boundary Street and £84 3s. for the self-contained tenements designed by Plumbe for Section E. The average size of living rooms in the Trust dwellings was larger at 176 sq. ft, but they were built in seven-storey blocks compared with the Council's five, with a less than 45 degree angle of light to the lower rooms and deficient through ventilation. It was claimed that these differences reduced the costs of construction per room. A later report described the Guinness Buildings as built on an associated plan, with one WC and one sink to four tenements being the usual proportion. The standard of finish was, however, comparable to that in Council buildings.[24]

The later buildings on the Boundary Street estate, which usually contained a mixture of self-contained flats and flats with detached but private sculleries, mostly ranged between £90 and £100 per room. All these had a minimum of 160 sq. ft living room compared with 146 sq. ft in Streatley Buildings. The later reduction of room sizes and reintroduction of the balcony principle reduced costs in a period of rising prices, but the Duke Street dwellings still required £86 per room. With reductions in the standard of finish the Cotton Street dwellings were, however, finished in 1901 at £71 per room. When this standard was in turn abandoned in favour of that at Swan Lane costs became £88 per room, and the dwellings on the Council's last clearance sites were completed at about that cost.

Although the building costs per room of the Council and the Trusts do not seem to be very far apart, costs quoted by other companies were often lower. The Improved Industrial Dwelling Company secretary in 1882 estimated the building costs on its last four estates at £53–55 per room. These were for self-contained tenements. However, IIDC costs were later reckoned to have risen from £40 per room in 1863 to £75 in 1900.[25] More information is available concerning the East End Dwelling Company's venture at Katherine Buildings, Rosemary Lane, built in 1885. The buildings comprised a five-storey block with balconies at the rear giving access via short passages to groups of five rooms. With no water supply to the rooms, and sinks and WCs grouped in the passages, the building cost amounted to £57 per room.[26] Blocks built by Liverpool Corporation in the 1880s had building costs of £81–91 per room. In 1897, however, Gildart's Gardens was completed at an estimated £36–44 per room, so triggering the national debate on building standards. In these three-storey tenements much of the exceptional cheapness was due to the use of old materials and standard of finish. But no other blocks were built to this standard; and in the second phase of building at Gildart's Gardens, completed in 1904, the building cost per room was £70. This was about the average for later Liverpool building, which does not, therefore, appear in so dramatic a contrast with current LCC construction.[27]

In addition to the moral and sanitary considerations mentioned in the first section, two economic objections were made against lower building standards. Where this was achieved by reduction in the standard of finish and quality of materials, it was argued that immediate cost reductions would be achieved at the expense of the future. This point was strongly put by the LCC architect in his comments on Gildart's Gardens. He considered it 'inexpedient for any public authority to build on such a poor standard, outgoings for repairs and maintenance will probably be very heavy in future, and the appearance of these buildings conveys the impression that they have been built twenty years instead of four'.[28] Where lower standards were achieved in design then the kind of argument used by Waterlow came into force: 'It would not have been right to build for the lowest class, because you must have built a class of tenement which I hope none of them would be satisfied with at the end of fifty years.'[29] This was also the defence of the Council for their building regulations.

One argument for housing subsidies is that the market attaches the nature of a building with many years' durability to the conditions of the moment. This may lead to undesirable conditions being built in which are later difficult to eradicate, as Cross's Act itself emphasized. Alternatively, in certain cases basic structural conditions may be difficult to alter in line with rising expectations. In Cross's Act, however, the notion of subsidy was tied to the removal of the mistakes of the past. There was, in theory at least, no subsidy related to the present or future which would carry with it the implication that tenants were being provided with better housing than their station in life would otherwise permit.

In one way it mattered little whether a subsidy was attached to land or to buildings. Either way the overall cost would be reduced and a lower rent could be demanded. It did make a difference, however, in the comparative advantage of one site compared to another, and in relating operations under Cross's Act to other forms of housing action. The advantage to the tenant of relatively cheap sites, such as that of the former Caledonian Asylum, was lost when the ground rent per room was similar to that at Boundary Street. Indeed, partly because of lower densities, even on the Council's Part III suburban estates land costs were of the order of £15–17 per room. Other effects were seen in the field of rehabilitation. When the Council investigated the possibility of buying subdivided middle-class houses and improving them by the addition of a common washhouse, separate WCs for each family, and a water supply on each floor, it was found impossible to charge rents lower than in its newly built tenements.[30]

Rents and tenants

Although there was obviously a relationship between costs of construction and rents, it was not an automatic one. This was particularly true in the case of the

Peabody Trust, which had relatively high construction costs but offered rents lower than those of other dwellings companies. MADIC accused the trustees of being 'unfair traders' who operated 'at least 30 per cent under the market'.[31] Peabody practice made it possible for them to offer rents per room on the Cross sites which were on average 4 per cent below those prevailing before clearance. Making allowance for the different mix of tenement sizes, there was probably a slight increase, and although on some formerly high price sites such as Bedford-bury rents were much reduced, at Rosemary Lane Whitechapel, where the average rent per room before clearance had been 2s. 1d., rents were afterwards 3s. for one room, 4s. 9d. for two, and 5s. 9d. for three.

It is known that only 11 of the old tenants entered the new Peabody Buildings at Whitechapel. The composition of the new tenants can, however, be examined from the Booth schedules and is summarized in Table 8.1. Overall, only three of Booth's classes were represented, with a dominance of E (56 per cent), together with D (38 per cent) and B (6 per cent). This composition varied little according to tenement size, and there was actually a greater proportion of Classes B and D in the three-roomed tenements than in the two-roomed ones. Labourers, particularly those in the higher supervising and portering jobs, formed the largest of the labour sections (40 per cent). The 'other wage' group (26 per cent) formed the next largest section, one not much represented in the former population. Amongst them were a striking number of policemen. Artisans and female-headed households were also represented, but there were few traders. The blocks as a whole were classified as purple, or mixed social composition.

The Peabody operations reached further into the working classes than those of most dwellings companies. There was no representation at Whitechapel of Booth's Class F – earning over 30s. a week – which Hobsbawm once identified with the 'labour aristocracy'.[32] Tenants had had little time to increase their earnings above the 25s. limit for first entrants. On the six succeeding Cross sites

Table 8.1 Tenant composition at Peabody and Katherine Buildings, Whitechapel, 1887.

Booth industry section	Tenants		Booth social class (%)		
	No.	%	B	D	E
Peabody Buildings					
labour	83	40	6	46	48
artisan	39	19	2	26	72
other wage	54	26	2	24	74
female	20	10	30	55	15
other	9	4	—	56	44
total	205		6	38	56
Katherine Buildings					
labour	62	64	26	40	34
artisan	9	9	11	44	44
other wage	6	6	17	17	67
female	16	16	44	31	25
other	4	4	25	50	25
total	97		26	38	35

Source: Booth notebooks, British Library of Political and Economic Science.

Table 8.2   Tenant composition in London blocks of model dwellings, 1887–9.

Block type	Total tenements* (no.)	Booth social class (%)						
		A	B	C	D	E	F	GH
Group I	14 288	—	4	3	16	62	14	1
Group II	13 491	0	10	12	22	44	11	1
Group III	7 669	1	17	12	27	34	9	0
total	35 448	0	9	9	21	49	12	1

Source: Booth, C. 1889–1903. *The life and labour of the people in London*, 1902–3 edn, vol. III, pp. 9–14.

Notes:
Group I   Philanthropic and major companies.
Group II   Larger trading companies and private owners.
Group III   Smaller private owners.
Total includes a small number of tenements erected by employers for their workpeople.

*Tenements.

taken by other companies and individuals the rents were all much higher than Peabody rents, taking into account location, and they represented an average 34 per cent increase on the rent per room before clearance. On these sites the new population must have been much closer in composition to that established by Arkell for trusts and major dwellings companies – his Group I (Table 8.2). The main result of the Arkell inquiry was to establish a close relationship between the ownership of blocks, their sanitary character, and the composition of their inhabitants. His Group I contained 'nearly all the best blocks', and in Group II – the larger trading companies – buildings though usually sanitary were 'darker and more densely populated'. Group III, the smaller blocks of private builders, were 'dark and insanitary, constructed with little or no regard to the comfort of the occupants'. Thus, 'with a few exceptions it is only in the worst blocks that the poor are accommodated'.[33] The universe of block dwellings was thus not confined to 'respectable' owners and tenants but was seen to replicate the housing market generally. Only the weight of the various categories was changed.

Fortunately, the East End Dwellings Company block at Katherine Buildings was also included in the Booth notebooks. The 97 tenants recorded were divided into the same three classes as at Peabody Buildings, but much more evenly: 26 per cent were in Class B, 38 per cent in Class D, and 35 per cent in Class E (Table 8.1). As in Peabody Buildings, female-headed households were particularly concentrated in Class B. Otherwise Katherine Buildings had a population of 'labourers' without the large 'other wage' category of Peabody Buildings. Most of the rooms about this time were let single at 3s to 3s. 6d. or in pairs at 5s. to 5s. 6d., although some were cheaper. Despite the much cheaper construction of this block, therefore, rents were not less than those charged by Peabody, and the difference in tenant composition resulted primarily from management practice.

Bodies such as the Peabody Trust, the Guinness Trust, and the Council enforced strict rules regarding the occupation of rooms, limiting this to a maximum of two persons. As the work of these bodies was primarily aimed at the family, the one-room tenement was necessarily of little interest, whereas families of modest size by the standard of the time would require three rooms. At Katherine Buildings, however, although rules about overcrowding were enforced, these were less strict, allowing the occupation of single rooms by families. In January 1886, Miss Pycroft, the lady manager, claimed a population per room averaging 2.5, and in October the population of the buildings was 655 or 2.3 per room.[34] These figures were not much lower than the average at Rosemary Lane before clearance (2.6) and higher than that at Boundary Street (2.2).

Octavia Hill told the 1882 Committee that overcrowding regulations were partly designed to confine access to the better class of tenants: 'It is not the rentals that prevent them being taken into model dwellings, but no one is inclined to build single rooms because of the destructiveness of the lower class of tenant, and the difficulties of collecting their rent.'[35] But such factors were reflected in the net rental, and any large increase in the turnover of occupants made it difficult to vet prospective new tenants in the usual way without increasing the loss on empties. Such difficulties certainly arose at Katherine Buildings in the early years; indeed it appears from the book of tenants that 78 per cent of the original inhabitants of 1885 had left by the end of 1886. Although many tenement blocks were to have high turnover rates, including those of the Council, this was exceptional. In the first ten months 27 tenants were evicted for bad conduct or non-payment of rents, and 56 left of their own accord, some to nearby blocks. In January 1887 Pycroft recorded: 'It has been disappointing that tenants of the lowest class cannot be kept in the buildings if order is to be preserved, and I have been obliged to turn out the rough set.' In this selective process the occupation rate was reduced to 1.95 persons per room in October 1889.[36] It should be noted that the Booth survey must have been taken after much of the initial turnover had taken place.

There can be no doubt that these bruising events had an important influence. Beatrice Potter (Webb), assistant to Miss Pycroft, recorded in her diary of December 1885: 'These buildings are to my mind an utter failure... the respectable tenants keep rigidly to themselves... Boys and girls crowd on the landings – they are the only lighted places in the buildings – to gamble and flirt. The lady collectors are an altogether superficial thing.'[37] Octavia Hill continued to believe that life in the blocks managed 'under rules they grow to think natural and reasonable, inspected and disciplined, every inhabitant registered and known, school board laws, sanitary laws, and laws of landlord or company regularly enforced' was the future of the 'respectable working man'. But she now also feared 'the massing together of herds of untrained people' in blocks where 'the low class people herd on the staircases and corrupt one another, where those a little higher would withdraw into their little sanctum'.[38] Her advocacy of retaining slum tenants in their existing dwellings dates from this period. The return of the Home Office and Council to stricter standards in the early 1890s may also have owed something to this experience.

At Boundary Street there was a considerable range of rents, varying according to the cost and quality of the accommodation. The few one-room flats were priced at 3s. 6d. Two-room flats ranged from 5s. 6d. in Culham and Sonning Buildings to between 6s. 6d. and 7s. 6d. in the later blocks. Similarly, the price of three rooms ranged from 7s. 6d. to between 8s. 6d. and 9s. 6d. If, however, we

Table 8.3 Boundary Street, 1892–5: rents of existing tenants related to minimum Council rents. The table shows the percentage of households in various size groups able to afford a given number of Council rooms from their existing pre-clearance rents. Minimum Council rents are taken: 3s. 6d. for one room, 5s. 6d. for two, 7s. 6d. for three, and 9s. 6d. for four.

Household size	No. of tenants	Rooms obtainable				
		Nil	One	Two	Three	Four
1–2	152	58	30	7	3	2
3–4	138	33	46	12	6	4
5–6	124	19	53	10	11	7
7–8	62	10	45	27	10	8
total	476	34	43	12	7	5

Source: LCC HC Presented Papers, Boundary Street tenant compensation schedules.

take the minimum rents for each room size as an indication of the accommodation the Council could reasonably have provided, it is possible from the data considered in Chapter 7 to calculate the proportion of tenants who could have taken tenements of various sizes without increase in their existing rents. Overall, some 34 per cent could not have afforded even a single room, and another 43 per cent one room only (Table 8.3).

These calculations can be taken further by also considering household size in relation to the overcrowding regulations of the Council, which imposed a maximum of two persons per room. On this basis, 25 per cent of existing tenants could have afforded Council accommodation. The largest group of these were, however, one- and two-person households occupying a single room – non-family groups for which Council building was not intended. Moreover, two other considerations reinforce a pessimistic view. The figure could not be changed dramatically by marginal moves: reducing minimum rents for all tenements by 6d. a week would have added another 7 per cent potential uptake, whereas allowing three persons per room would have added a further 9 per cent. Above all, the statistics only relate to those who received compensation, and were probably of above average standing, and they include tenants who met their rental commitments by subletting (see Ch. 7).

Like the dwellings companies, the Council exercised care in the selection of its tenants. They had to complete an application form giving family numbers and ages. Superintendents of dwellings were instructed that on receipt of the form 'a personal visit is to be made . . . to the applicant's address and a report made . . . as to the rent paid, condition of rooms, and the desirableness or otherwise of the applicant. No application is to be accepted unless the valuer's written authority is received.'[39] If childbearing subsequently brought the family above overcrowding levels, they were allowed to remain in their existing flat until the children were two years old. Later, in March 1900, the overcrowding rules were relaxed, so that children under five were not counted and children from five to ten were counted as half.[40]

When, therefore on the opening of Streatley Buildings the valuer wrote to 140 Boundary Street tenants inviting them to make application, they were certainly already a selected group. It is known that 56 declined on the ground that rents were too high. Other reasons given for refusing the accommodation included objections to model dwellings themselves and their appearance. The most important ground, other than rent, was inconvenience for workshop or trading accommodation – 'wants large shop', 'wants yard', 'wants place for barrows', 'workshops not large enough' were typical responses.[41] As those with workshops and shops, and the more prosperous dealers, were the class of population at Boundary Street best able to afford the new rents, their rejection of the new accommodation was particularly disappointing. The Council had included detached workshop accommodation in the Streatley Buildings complex, but this arrangement, and the prospect of greater regulation, was not acceptable in most cases.

The experience of the Council thus largely replicated that of the dwellings companies, and although after Boundary Street the trend of opinion swung back once more in favour of building down, very little new was added. More significant was the contrast with the Council's plans for rehousing adopted in 1890. These had envisaged clearance in sections so that 'when the first section is cleared and artisans' buildings are ready, the inhabitants of the second section will be moved into the first section'. The initial displacement was to be met by building at Goldsmiths Row, Shoreditch, for 500 persons, but later the site proved impracticable for block dwellings. It was also hoped to rehouse 480 in the Guinness Trust buildings in Columbia Road, but it was found that applications from Boundary Street tenants were 'not encouraged'.[42] In the end all of these hopes and plans came to nothing, and although the Council began a record of its new tenants in Streatley Buildings, only this initial list has survived. Ten tenants from the scheme appear upon it.[43]

Like the dwellings companies before it, the Council aimed to make a difference by increased social responsibility, tighter management, and greater economic efficiency. These factors were important, and they helped some groups, but they could have only a marginal impact on the character of the housing market, even with the land subsidy under Cross's legislation. There was some choice in the matter of rents, tenants, and building standards around which policies could oscillate, but it operated within strict limits. The Council's dual rôle as guardian of the sanitary regulations and as market trader made the dispersal of most of the existing tenants the only logical outcome of slum clearance operations.

Clearance and dispersal

In the debate on Cross's Act Shaftesbury had emphasized that demolitions which 'may ultimately produce great good' had at least severe short-term effects. The populations involved were entirely absorbed in day-to-day living.

> And so when the moment arrives . . . they are like persons possessed – perplexity and dismay are everywhere; the district has all the air of a town taken by assault. Then they rush into every hole . . . all struggle to be as near as possible to their former dwellings. Streets and houses already overcrowded have doubled in population. Every demand of health and decency is set aside.[44]

It was because of such argument that provision had been made for the larger schemes to be cleared in stages, and pursuing this line of thought, in February 1879 the Home Office refused to allow the Board to clear any more property until progress had been made in erecting houses on cleared sites.

The Board immediately highlighted the effects of planning blight on properties for which it was now seen to be responsible, and in many cases had become the actual owner. It argued that some dispersal was taking place anyway as the worst houses had to be closed, and others became vacant as tenants moved away. Such houses 'became a great source of nuisance to the surrounding neighbourhood', and were 'the means of harbouring thieves and other similar classes'.[45] Support was received from the vestries and their medical officers, and Whitechapel wrote to the Board to complain that 'rebuilding dwellings in sections delays the improvements and from loss of interest vastly increases the cost'.[46] The Board's architect nonetheless argued for the maintenance of houses and a limited programme of repairs fully covered by the rental: 'The misery and privation they have to endure . . . are intensified by their compulsory removal, even when the places they occupy are little better than hovels'.[47]

The Board, however, was fearful of the possibility of scandal and anxious to be relieved of its rôle as slum property owner as soon as possible. Above all stood the view that the Board 'after paying large sums of money for these premises' should be able to 'clear the sites and realize some of their expenditure'.[48] Once the Peabody deal had been effected, and new buildings quickly erected on the Whitechapel site, the Home Office ban was lifted on the advice of the Select Committee. Staged clearance then next made its appearance at Boundary Street, but once it became clear that existing tenants would not enter the new buildings, the remaining parts were cleared fairly quickly and without any real co-ordination with the reconstruction side. In the major schemes of the turn of the century, however, staged clearance was again insisted upon by the Home Office, in the context of contemporary concern about overcrowding.

The other method developed to control dispersal was through the system of compensation payments. This had probably been used even in the schemes which Shaftesbury described and was certainly used by the Board. The extent of control is difficult to gauge. The Council would not have approved of the Board's practice of paying tenant compensation at Whitechapel in the White Swan public house. The Council valuer described his practice in 1892. An office was set up at Boundary Street which had information on local vacancies. Along with 13 weeks' notice to quit, tenants received a circular stating: 'As soon as you have obtained acccommodation elsewhere, I shall be prepared to make you an allowance to cover your loss by removal. When you have found another place, please send me the address, number of rooms and rent asked.' When this was done, the new premises 'were inspected by the valuer, and if there were any doubt as to their fitness for habitation, the medical officer was called in and his concurrence obtained before allowing any compensation'.[49]

Contemporary testimony concerning the dispersal of tenants was highly variable and obviously influenced by political considerations. There was a marked contrast, for instance, between the balance of evidence given to the Select Committee and that given to the Royal Commission. The former was dominated by medical officers who sought to deny the existence of any important problem. At Whitechapel, according to Liddle, 'they have not gone into the new buildings nor have they gone into any portion of our district that I am aware of.'[50] The Royal Commission, on the contrary, reported that clearances had led to increased rents and overcrowding in the locality, but the quality of its evidence was not much more satisfactory. Drawn from a different set of witnesses, including many clergymen, it is useful in a general way but offers few precise leads. For example, the Report claimed that in St Lukes, Clerkenwell, 'the district has never yet recovered from the pressure which was caused by the pulling down . . . for Peabody Town', and to support this referred specifically to the evidence of the Revd George Smith. Smith's view was, however, that although the demolition had created overcrowding, not only by the outflow of tenants but also the later inflow of building workers, once the new buildings had been constructed this special overcrowding had been 'quite relieved'.[51]

The compensation statements at Boundary Street provide a certain amount of information on the initial relocation of tenants. The data which the Council itself collected, to midsummer 1895, recorded that of 1035 tenants whose movements were known, 56 per cent had moved within a quarter of a mile of the site, a further 30 per cent within a quarter of a mile to half a mile, 9 per cent between half a mile and one mile, and 5 per cent over one mile.[52] A map of the dispersal, covering 537 tenants whose addresses are known, shows that the area immediately east of the site received the greatest number, whereas among those who moved further the predominant tendency was outwards towards the north and east, a few reaching as far as Dalston and Hackney Wick (Fig. 8.1).

Data on post-clearance rents were collected only for the first 159 tenants. As these sometimes changed accommodation, or even household composition, the best way of presenting the change is to examine the new rent structure which they were confronted with compared with the old (Table 8.4). Overall, there is not only the expected overlap between rents within the scheme area and without but also the expected increase in average levels, amounting to 7 per cent for single rooms and 16 per cent for two-roomed tenements.

Booth observed that although in many districts the population was always on the move, usually 'the people do not go far, and often cling from generation to generation to one vicinity, almost as if the set of streets which lie there were an isolated country village'.[53] A report of 1894 remarked that the population of Bethnal Green was particularly 'clannish . . . this characteristic arises from the fact that nearly all the people earn their livelihood without going out of the district. Bethnal Green is a parish of home industries.' It was specifically noted regarding Boundary Street that the people clung to 'these insanitary quarters uncomplainedly as indispensable to their daily work'.[54] According to the compensation statements, in 1892 when clearance began, 42 per cent of the tenants recorded had been living in the area less than 5 years, 28 per cent 5 to 9 years, 11 per cent 10 to 14 years, 8 per cent 15 to 19 years, and 11 per cent 20 or more. However accurate these figures are, there was certainly an important group of 'stayers', and many highly rented tenants chose to relocate relatively close to the site. In general, too, the pattern of the dispersal recorded in Figure 8.1 appears to be related to the contemporary location of the furniture trade.[55]

The only contemporary inquiry into links between employment and residence involved 3170 persons at Clare Market in 1896. It began by setting aside 544

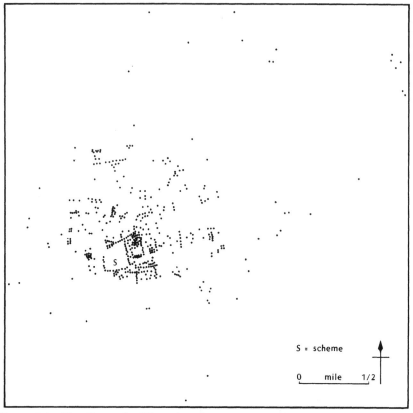

Figure 8.1 The dispersal of Boundary Street tenants. (*Source:* LCC HC Presented Papers, Boundary Street tenant compensation schedules.)

persons who were in common lodging houses, unemployed or dependent on relief, as having no necessary connection with the area. Of the rest, 1129 had employment in the 'locality' – 504 with exceptional hours (many at Covent Garden), 403 in other fixed employment, 99 hawkers, and 123 casual workers. The remaining 1300 worked beyond, of which 1083 had fixed employment and 657 of these worked within one mile. From this survey the Council decided that rehousing for 500 should be constructed on the site.[56] Later in 1899 a figure of 744 persons dependent on work with abnormal hours was given, and this was the origin of Thompson's claim that 'of the 6000 persons' in the Kingsway and Clare Market schemes 'only 774 belonged to a class whose work required residence on the spot a necessity'.[57] In 1907 Parsons was emphasizing that 'many of those who are living in crowded conditions in central districts cannot... show reasons... connected with employment. They are attached to the spot or the neighbourhood by habit, the proximity of friends or the allurement of endowed or voluntary charity.'[58]

Booth believed that the population of the new buildings was drawn to some extent from the wealthier groups in neighbouring localities. Other proponents

Table 8.4 Boundary Street tenants, 1892–5: pre- and post-clearance rentals.

One-room tenements

	No.	2s.	2s. 6d.	3s.	3s. 6d.	4s.	4s. 6d.	Mean
				Weekly rents (%)				
before	51	18	27	24	16	12	4	2.94s.
after	43	7	30	19	21	16	7	3.14s.

Two-room tenements

	No.	3s. 6d.	4s.	4s. 6d.	5s.	5s. 6d.	6s.+	Mean
				Weekly rents (%)★				
before	52	8	25	27	13	17	10	4.65s.
after	50	4	14	10	12	22	38	5.39s.

Source: LCC HC Presented Papers, Boundary Street tenant compensation schedules.

★Up to and including the rent stated.

of a suburban solution, like Masterman, disagreed, arguing that they were not inhabited 'by any of the people from the neighbourhood . . . by introducing a fresh set of people from outside you thereby add still further to the overcrowded character of the district'.[59] As usual, few of these views were based on any attempt to collect reliable evidence, and this is now difficult to obtain. At Streatley Buildings, Boundary Street, of the 43 tenants recorded from outside the scheme, 35 per cent came from within a quarter of a mile, and a further 11 per cent from within half a mile, with 25 per cent coming from over a mile. There was a predominant inward movement, but with few from distant suburbs. Llewellyn Smith examined the last addresses of Peabody tenants at Shadwell and Commercial Street, 'the great majority of them being in the immediately adjacent districts'.[60] On the Council's Millbank estate in 1902, 588 out of 896 tenants had previously lived in Westminster. However, a large proportion of the remainder had moved in from suburbs further south, including 97 from Lambeth and 36 from Battersea.[61]

No contemporary attempt was made to follow up the condition of the displaced Boundary Street tenants after their place of initial relocation was recorded, and of course it is not yet possible to attempt this from census enumerators' schedules. The clearance and rebuilding of Boundary Street did, however, lie between the first and revised versions of the Booth street maps, so that these provide some record of change. Describing this, Llewellyn Smith says, 'Church Street immediately south of the cleared area had changed from pink to purple with a black stripe, and black or black stripes had appeared in a number of streets in the vicinity previously free from them (e.g. Ebor Street, Chance Street, Princes Street, Chambord Street and Newling Street.'[62] Booth's own view, too, was: 'Everywhere these people are recognized as coming from the Nichol, and everywhere they have brought poverty, dirt and disorder with them, and an increase in crowding.'[63] As they moved in, he believed more existing residents tended to move out, so causing a population turnover which lowered the social condition of the streets affected.

Although such population movements seem very likely, another reading of the Booth street maps might emphasize the stability of social patterns in view of the major upheaval which had intervened. The main reception areas for Boundary Street tenants were already poor streets, many of them coloured dark blue, and it is their geography that determines the local patterns recorded in Figure 8.1. The Booth survey, it must be remembered, was prone to award an additional black stripe on the basis of small numbers of people. Although the number of Boundary Street tenants was large, the streets around were also very populous. Even within a quarter mile radius, at prevailing densities, there would have been 25 000 residents, and there would have been 95 000 within half a mile. Over half a mile north of Boundary Street lay Wilmers Gardens 'with thieves, prostitutes, bullies and flower sellers and cadgers, having received incomers from Boundary Street'.[64] Although it is certainly true that the longer-distance movements of Boundary Street tenants also often involved poor and deteriorating streets, only two are recorded as having moved to Wilmers Gardens.

Clearance and city structure

There is certainly a danger in considering patterns of population change in relation to schemes like Boundary Street that all new developments, whether for good or ill, will be ascribed to the scheme itself. Despite the magnitude of population displacement involved, however, there were other contemporary population changes of some importance taking place. During the 1890s the Jewish population had expanded north from Whitechapel across the Bethnal Green Road to Ebor Street, Chance Street, Church Street, and blocks in Newling Street, Chambord Street, Gossett Street, and the Boundary Street, estate itself.[65] More generally, all the local developments regarding schemes have to be placed in the context of the major population changes occurring in London as a whole. The maps of population density and change prepared by Grytzell remain the best depiction of this process.[66] By 1901 the characteristic pattern of low-density core succeeded by a sharp transition to a belt of high density, which in turn gives way to the successively lower densities of the suburbs, was strikingly apparent. Over a wide area population was declining, and population movements must have shown a marked net outward tendency, allowing for natural increase. In 1891–1901 all parts of Shoreditch were recording slight net losses, whereas Bethnal Green North and South were recording slight net increases, in common with neighbouring parts of Whitechapel.

The main measure of the balance between population change and housing stock in the 19th century was the index of persons per inhabited dwelling. Between 1871 and 1911 this declined only in Westminster and the City, but there were increases of at least 20 per cent in Finsbury, Shoreditch, Bethnal Green, Stepney, and Southwark. However, due to the extension of flats and business premises partly used as dwellings, by 1911 the proportion of the population living in ordinary dwelling houses was under 50 per cent in Westminster and under 70 per cent in six other boroughs. As a block of flats was counted as a house, this inflated the figures for many districts, although inhabited business premises usually reduced it. A second index of persons per room which avoids many of these problems was not available until 1891 and then only for one-to-four-room tenements. The intercensal results later available contradicted those obtained from the persons per dwelling index. In London as a whole persons per house increased from 7.73 to 7.93 between 1891 and 1901, falling back to 7.89 in 1911. Over the same period the persons per room index moved from 1.58 to 1.48 and then to 1.42.

In the ten central boroughs of Bermondsey, Bethnal Green, Chelsea, Finsbury, Holborn, St Marylebone, Shoreditch, Stepney, Southwark, and Westminster, the net loss of population between 1871 and 1911 amounted to 201 000. The number of inhabited dwellings recorded a fall of 42 500. At about six rooms per dwelling (near the average in 1911) this would represent some 250 000 rooms. In the same area slum clearance schemes removed about 18 000 rooms (gross) during this period. At the most they accounted for 10 per cent of demolitions, and even allowing for street improvements, schools, and other factors it is clear that the overwhelming cause of loss of rooms must have been demolition or conversion for commercial purposes.

It is possible to examine the impact of slum clearance more closely within the borough of Finsbury.[67] This was one of the areas most affected, with clearances in Clerkenwell and St Lukes involving 6623 people and demolition of 3176 rooms. Over the period between 1871 and 1891 the population fell from 124 722 to 87 923, and the number of inhabited dwellings by 5515, probably representing some 30 000 rooms. Data collected by Newman between 1902 and 1905, when the Aylesbury Court and Garden Row clearances were taking place, suggest that even within this concentrated period the LCC schemes accounted for only 36 per cent of the population displaced by closure and demolition. Newman also prepared a record of model dwellings in Finsbury, these housing a population of 17 130 in 1901. The largest number, housing 8410 persons, were built in the 1880s and include 4118 persons at Pear Tree Court and Whitecross Street. Neither in dehousing or in rehousing did clearance schemes have any clear field of local predominance through which their medium-term effects could be isolated and accurately estimated.

Reviewing all this information, Newman concluded: 'Finsbury is at present in the transition stage from a residential district into a non-residential commercial district comparable to the City of London. Such a transition acts prejudicially for a long period of years on dwelling house property and overcrowding.'[68] He took an optimistic view, claiming that in Finsbury the prejudicial period was now drawing to a close, whatever might yet be to come further out. Such observations reflected the thinking of their time regarding the 'normal' development of the city. The Cross strategy, by contrast, had conceived of slum clearance as a set of local operations clearing up the mistakes of the past. It had clearly not been designed to extend central commercial functions; indeed in so far as it reserved land for working-class occupation it might be regarded as a reaction against this. Particularly for those in the Shaftesbury tradition, commercial expansion might be seen as a cause of social problems and a contributory factor to the slum. By the end of the period, the suggested levy on private developers in the form of a purchase of sites fund strengthened this tendency, as did the renewed emphasis on housing operations. By contrast, a suburban solution certainly involved no restriction of central business expansion, and indeed the two would seem to be positively related.

In some ways, however, slum clearance and central business expansion were complementary. The latter was, after all, a principal means by which slums and their populations might be removed under the market. Already in the 1860s, when Shaftesbury was emphasizing a local shifting of the slum from one spot to another, others like Denton were highlighting a more general movement and accusing the City Corporation of having 'shovelled out the poor in order to lower the poor rates of the City parishes.'[69] From this point of view, the removal of the slum by ordinary commercial processes was slow and uneven, leaving even in its later stages densely inhabited pockets of survival. Slums were being squeezed out of Holborn and Westminster in the 1840s if not before, but the process was still not completed at the end of the century. Nonetheless, there were clear signs that the social character of these areas was being changed. Using the statistics which Llewellyn Smith calculated retrospectively for the percentage of overcrowding in all tenements, an index of 1911 based on 100 for 1891 shows figures of only 56 for the City, 60 for Westminster, 61 for Holborn, and 73 for St Marylebone. Such index figures contrast with those of 103 for Stepney, 101 for Shoreditch, 100 for Bermondsey, 95 for Finsbury, 94 for Bethnal Green, and 92 for Southwark.[70]

It is useful to consider these trends in relation to the first general survey of unfit housing in London carried out in 1911.[71] Of the 19 678 houses represented on the priority list, 10 516, or 53 per cent, were located in the ten central boroughs previously mentioned, which had included 76 per cent of the rooms demolished in Victorian slum clearance. Within this group, however, there were some notable differences. Holborn and Westminster, which had accounted for 17 per cent of Victorian slum clearance, now represented only 542 houses, or under 3 per cent of the total. The six boroughs surrounding the City to the north, east, and south – Bermondsey, Bethnal Green, Finsbury, Shoreditch, Stepney, and Southwark – represented 9212 or 47 per cent, and had accounted for 56 per cent of Victorian clearance.

Cross's Act had envisaged the removal of slum blocks conceived as an historical legacy. Indeed, Cross himself claimed in 1900: 'I do not believe that there is one of these unhealthy areas now existing in London that it was the special object of the bill of 1875 to remove.'[72] At best this claim could be sustained only in relation to a central core, although that part was, of course, of special political significance. There, clearance schemes were undertaken at a relatively late stage of 'transition', mopping up areas that were consequently costly to deal with per person but of relatively limited extent. A theory of urban transition, however, implied that the slum was a normal feature of the urban structure, which if tackled predominantly by clearance required a large and continuing programme more remote from the centres of power. In the context of the period, such clearance would either require building many blocks of cheap tenements at high densities or further outward dispersal of large numbers of the poor, and in either case would be expensive.

Nor was it easy to combine clearance with other local operations given the emphasis on the clear boundaries of the slum, and in the wider world there was no monitoring, let alone control of land-use changes, which might have allowed a more flexible combination of clearance and rehousing, nor were there any subsidies or special compensation arrangements. Even the policy of ensuring adequate conversion of middle-class houses as they became subdivided faltered on this hurdle. It was in these circumstances that support for the continuance of clearance policies came to rely so much on their negative value as an alternative to any breakthrough of municipal building into the general housing market.

Like slum clearance, the Part III strategy also involved intervention on a narrow geographical front, and the mechanisms for transmitting beneficial effects back to the rest of the city, and particularly to its slums, were of doubtful efficacy. Even if large-scale expansion could be promoted, the resulting devaluation of buildings, in a context in which land values were still supported by corresponding expansion of city-centre functions, posed clear problems for slum formation. However, these potential difficulties had yet to be faced, whereas the defects of the existing clearance strategy had been hammered home over a long period. In terms of potential costs a suburban solution fitted more easily into the now familiar theory of the 'normal' development of the city, of which it was both product and parent.

Notes

1   Tarn, J. 1973. *Five per cent philanthropy*. Cambridge: Cambridge University Press; Beattie, S. 1980. *A revolution in London housing*. London: Architectural Press. More financial information is given in Hole, W. V. 1965. The housing of the working classes in Britain 1850–1914. Unpublished PhD. thesis, University of London.
2   PRO HO 45 10198/B 31375, pp. 45–8.
3   Sykes, J. 1901. The results of state, municipal and organised private action on the housing of the working classes. *Journal of the Royal Statistical Society* Vol. LXIV, pp. 212–3.
4   MBW Minutes WGP 19 Nov. 1877 (36).
5   MBW 2411/7 Reports nos. 727 and 732; Minutes WGP 3 July 1876 (55).
6   MBW 1878 Whitecross Street Local Inquiry, 14 April 1877.
7   MBW Minutes WGP 24 Feb. 1879 (60).
8   MBW Minutes WGP 8 March 1883 (13), 2 March 1885 (75).
9   LCC Minutes 3 Dec. 1889 (11).
10 LCC Minutes HC 13 Aug. 91 (24).
11 Fleming, O. 1900. Working class dwellings: the rebuilding of the Boundary Street estate. *Journal of the Royal Institute of British Architects*, vol. VII, p. 6.
12 LCC HC Presented Papers 1898–1900, Bundle 11.
13 LCC HC Presented Papers 1898–1900, Bundle 33.
14 PRO HO 45 10198/B 31375, p. 42.
15 LCC HSG/GEN/2/2, Section II, no. 41.
16 MBW 2411/7 Report no. 908.
17 These and subsequent statistics are derived from LCC 1900. *The housing question in London*, *passim* and 294–320; *London Statistics*, 1897–8, vol. VIII, p. 270, 1905–6, vol. XVI, pp. 119–126, and 1912–13, vol. XXIII, pp. 204–5. The six non-Peabody schemes were at Rosemary Lane (part), St George the Martyr, High Street Islington, Bowmans Buildings, Essex Road, and Wells Street. Statistics are incomplete for Goulston Street. The Board schemes completed by the Council were at Brook Street, Trafalgar Road, Hughes Fields, and Cable Street. Shelton Street is excluded because of its common lodging house provision. Later Council schemes were at Churchway, Webber Road, Garden Row, and Aylesbury Court.
18 *Hansard*, vol. 222, HC Deb., 3s., 15 February 1875, col. 341.
19 LCC Minutes 18 July 1893 (16).
20 Newsholme, A. 1891. The vital statistics of Peabody buildings and other artisans' and labourers' block dwellings. *Journal of the Royal Statistical Society*, vol. LIV, pp. 91–2.

21   LCC HSG/GEN/2/2, Section II, no. 41.
22   ibid. and Newsholme, A., op. cit., pp. 91–2.
23   LCC Minutes HC 13 Dec. 1893 (13).
24   LCC HSG/GEN/2/2, Section I, no. 16, Section II, no. 41.
25   Select Committee on Artisans' and Labourers' Dwellings 1881–2, *PP* VII, 1881, 4793; Perks, S. *Residential flats of all classes*, p. 81. London: Batsford.
26   *The Builder*, 28 March 1885, pp. 446–63.
27   Thompson, W. 1907. *The housing handbook up to date*, pp. 44–7. London: National Housing Reform Council.
28   LCC HSG/GEN/2/2, Section IX, Report by Architect on Glasgow and Liverpool working class dwellings, 2 Oct. 1901.
29   Royal Commission on the Housing of the Working Classes, *PP* XXX, 1885, 11959.
30   LCC HSG/GEN/2/2, Section II, no.51.
31   Select Committee, op. cit., 4072.
32   Hobsbawm, E. J. 1964. *Labouring men*, p. 285. London: Weidenfeld.
33   Arkell, G. 1892. Blocks of model dwellings: statistics. In Booth, C. The *Life and labour of the people in London*, 1902–3 edn, vol. III, pp. 6–8 and 28–9.
34   Webb Collection, Coll. Misc. 43. Katherine Buildings: Record of the Inhabitants 1885–90. British Library of Political and Economic Science, London.
35   Select Committee, op. cit., *PP* VII, 1882, 3314.
36   Webb Collection, op. cit.
37   Webb, B. 1945. *My apprenticeship*, pp. 237–8. London: Longman Green.
38   Hill, O. 1892. Blocks of model dwellings: influence on character. In Booth, C., op. cit., pp. 32, 35.
39   LCC HSG/GEN/2/2, Section I, no. 18.
40   LCC Minutes HC 14 March 1900 (14).
41   LCC HC Presented Papers 1895–1896, Bundle A3, 22 May 1895.
42   LCC Minutes HC, north-east s/c, 7 Nov. 1892 (4).
43   LCC HC Presented Papers 1895–1896, Bundle A3, 3 May 1895.
44   *Hansard*, vol. 224, HL Deb., 3s., 11 May 1875, col. 454.
45   MBW Minutes WGP 7 March 1881 (16).
46   Select Committee, op. cit., *PP* VII, 1881, 432.
47   MBW 2411/7 Report no. 929.
48   MBW Minutes WGP 21 March 1881 (63).
49   LCC HC Presented Papers 1889–1892, Bundle A3, 13 Sept. 1892.
50   Select Committee, op. cit., *PP* VII, 1881, 286.
51   Royal Commission, op. cit., pp. 3679–3682.
52   LCC HC Presented Papers 1895–1896, Bundle A3, 9 July 1895.
53   Booth, C., op. cit., vol. 1, p. 27.
54   *Municipal Journal*, 15 June 1894, p. 310.
55   Compare Hall, P. 1962. *The industries of greater London since 1861*, p. 78 (Fig. 19). London: Hutchinson.
56   LCC HC Presented Papers 1896–1897, Bundle 41.
57   Thompson, W. 1903. *The housing handbook*, p. 50. London: National Housing Reform Council.
58   Parsons, J. 1903. *Housing by voluntary enterprise*, p. 53. London: King & Son.
59   Masterman, C. 1901. Reprint 1973. *The heart of the empire*, p. 102. London: Harvester Press.
60   Booth, C., op. cit., vol. 3, p. 80.
61   LCC Minutes HC 29 Oct. 1902 (18).
62   Smith, Sir H. L. 1932. *New survey of London life and labour*, vol. 3, pp. 140–1. London: P. S. King.
63   Booth, C., op. cit., Religious Influences Series, vol. 2, p. 71.
64   ibid., vol. 3, p. 115.
65   Royal Commission on Alien Immigration, *PP* IX, 1903, 6576.
66   Grytzell, K. G. 1956. *County of London population changes, 1801–1901*, p. 33. Lund University Studies in Geography Series B.
67   Newman, G. 1901. *Some notes on the housing question in Finsbury*. London: Bean.
68   Finsbury Borough Council 1901. Medical Officer of Health Report, p. 134.
69   Denton, W. 1861. *Observations on the displacement of the poor*, p. 23. London: Bell & Daldy.
70   Smith, Sir H. L. op. cit., vol.1, pp. 150–70.
71   LCC HC Presented Papers 14A–C, 13 Dec. 1911 (3 vols).
72   *Hansard*, vol. 86, HL Deb., 4s., 25 July 1900, col. 631.

9 Review

This investigation began from the position that Cross's legislation embodied a paradigm strategy for dealing with slums in late Victorian London. As such it was a problem-selecting as well as a problem-solving device; and its view of slum was shaped by the proposed remedies, just as those remedies were also shaped by conditions on the ground. By examining the manner in which this view of slum was formed, and its correspondence with conditions found on implementation of the legislation, it is possible to relate the selectiveness of the strategy to the consequences of its adoption. This final review sets out only to underline certain themes, and does not provide a summary of the argument, which needs to be set in detailed contexts.

It has been stressed that such a strategy was a complex, not a simple, unitary policy built around a single principle. It drew on several existing strands of thought and action, and new ones were later added as modifications were attempted in the light of experience. All these strands were deeply rooted, and each of them tells us something of the Victorian approach to the slum. In that sense, Cross's Act needs simply to be unravelled into its component parts. Yet the importance of the Act lay in its ability to weld these parts into a plausible coherence acceptable to dominant political opinion. It needed to reconcile the destruction of slums with acceptable compensation of property owners and sufficient reconstruction to avoid unpalatable overcrowding all at relatively low cost to the ratepayer. Evidently, it was the failure to achieve this reconciliation in practice that limited the application of the strategy. Nonetheless, this remained the most acceptable and most general framework from which action flowed in the late Victorian period.

A concentration on this framework thus highlights the variety of concerns which affected the approach to clearance. For if the fundamental drive was an attempt to shore up the base of society by imposing greater discipline – a drive in which slum clearance is linked through sanitary action to wider aspects of social policy – nevertheless this objective was cut across by others. The issues raised by overcrowding and property compensation, and the ambitions of philanthropic and municipal enterprise, all played an important part in the vicissitudes of the legislation. It was also from the reconciliation of various objectives that the exact form of demarcation of the slum resulted. This particular model then became part of the structure around which public debate ranged and around which politicians sought to weave an advantage. As in much relatively abstract discussion, a limited view of affairs was reinforced by constant feeding with one type of example.

Although the models of slum present in political strategies are not artificial concepts, they are unlikely to correspond very exactly with conditions on the ground. This is very apparent in a relatively crude case such as the Cross strategy. Here slum was separated from the rest of society in an emphatic manner and distributed in isolated blocks. True, an important theme was that these were bastions from which contagion spread, whether medically or socially; but they were also conceived as backwaters in which built environment and social behaviour had not been brought into the stream of modern life. From this stemmed an emphasis on the boundaries of such blocks, and a black uniformity which concealed a degree of overlap between conditions inside and outside, both in the nature of dwellings and social conditions. Beyond that came a lack of recognition of the true land-use character of slums, of the presence of industry, and of the high market value of slum property arising from its geographical situation in the city as a whole. Carried to extreme, the slum might even be seen as isolated from economic society and economic geography, an entity distinguished by moral rather than economic criteria. This was the irresponsible society which should be made to pay for the results of its own laxity and misconduct. The effect of these misunderstandings and oversimplifications was to push up costs, to reduce the scale of the clearance programme, and to shackle its effectiveness.

The treatment of property compensation is particularly interesting in showing how the same basic model of slum could be used in different political directions. Cross used it to lift the moral burden from landlords by pointing to conditions within slums which were beyond their control. But the model also had some initial advantages to the anti-landlord group. It highlighted the contrast between 'wretched hovels' and high compensation, and facilitated a transfer of moral stigma from tenants to owners. Later, however, the difficulties of applying one set of rules within small islands of property which were not being applied outside was one of the factors that defeated attempts at reform. This could not proceed beyond a certain stage without the achievement of wider change outside the slum.

Slum clearance as a public programme necessarily drew sharper attention to the consequences of demolition than any commercial operation. The responsibility of government was engaged at least to a certain degree, and this was reflected in Cross's Act with its insistence on the complementarity of its destructive and constructive parts. There was, however, some tendency to separate these parts by emphasizing the past in relation to the former and the future in relation to the latter. This may be inherent in any clearance strategy, particularly when attention is given mainly to buildings. But in the Cross strategy it was extended to the point that slum formation itself was seen as the product of a past society uncontrolled by current sanitary regulations. New construction, however, had to provide an acceptable environment for future bastions of sanitary and social order.

Such general issues were related more narrowly but more precisely in the financial equation of costs, rents, and subsidy. The cost of property compensation was crucial to the outcome, even more than building costs. If, as argued here, the reason for high costs lay in market values and in the failure to drive down those values through the concept of the unfit house, then the only way to effect reconstruction without large subsidy was to adopt the market solution – to displace the existing tenants and replace them with higher value users. The subsidy available in Cross's legislation altered the quantification of these variables, but left their fundamental import unchanged. There was some scope for oscillation between emphasis on building standards and emphasis on rents, but it was a limited one and not capable of a satisfactory equilibrium solution.

Linked to this were another set of problems relating to attempts to lift conditions in one department of life and not in others – quality or quantity, housing or other needs. The account of reconstruction and displacement given in the last chapter is thus largely a record of logical outcomes of factors previously discussed.

An understanding of the Cross strategy is enhanced by comparison with the alternative suburban strategy adopted by the London Progressives. This involved a new view of slum, but one which was still partial and selective, which also reflected the need to reconcile problem and remedy. Here, the slum still had to be demarcated because a focus of attention on unacceptable conditions was a necessary root of action. But the new strategy represented a shift away from an overwhelming concern with conditions at the base of society, in this respect forming part of a wider movement of social policy. It was an advantage to break down a narrow demarcation of 'slum', to identify the problem at a larger scale to which existing clearance policies could not be applied, and in some degree to acknowledge its relation to current economic and social processes. The slum was linked more closely with the rest of the urban structure, but analysis stopped short at the point where empirical referents were sufficient to support a plausible package at the political level. Critical attention was directed mainly at the defects of the existing policy.

From many points of view the new strategy appears as a breath of fresh air. It involved expansion of housing supply, release of municipal enterprise, benefit to those directly affected, and an avoidance of some of the worst features of a moral approach to the slum based on its separation from the rest of society. For all these advantages, it drew its main strength from turning aside from a route barred by apparently insuperable blockages. It had not confronted or overcome the problems of compensation and subsidy, income and housing or other needs, or beyond this another set of blockages in public attitudes to slums or slum dwellers. It was an oblique advance.

Index

Printed and bound in Great Britain by
TJ International Ltd, Padstow, Cornwall